The Meaning of Europe

The Meaning of Europe

Geography and Geopolitics

MICHAEL HEFFERNAN
University of Loughborough

A member of the Hodder Headline Group
LONDON • NEW YORK • SYDNEY • AUCKLAND

First published in Great Britain in 1998 by
Arnold, a member of the Hodder Headline Group,
338 Euston Road, London NW1 3BH

http://www.arnoldpublishers.com

Co-published in the United States of America by
Oxford University Press Inc.,
198 Madison Avenue, New York, NY 10016

British Library Cataloguing in Publication Data
A catalogue record for this book is available from the British Library

Library of Congress Cataloging-in-Publication Data
Heffernan, Michael J., 1959–
    The meaning of Europe : geography and geopolitics/Michael
Heffernan.
        p.   cm.
    Includes bibliographical references.
    ISBN 0-340-66189-5 (pbk.)
    1. Europe—Historical geography.   2. Geopolitics—Europe.
I. Title.
D21.5.H44   1998
911'.4—dc21   98-28789
                                                                                  CIP

ISBN 0 340 66189 5 (hb)
ISBN 0 340 58018 6 (pb)

1 2 3 4 5 6 7 8 9 10

Production Editor: Liz Gooster
Production Controller: Helen Whitehorn
Cover Design: Terry Griffiths

Composition in 10/12pt Palatino by Phoenix Photosetting, Chatham, Kent
Printed and bound in Great Britain by MPG Books, Bodmin, Cornwall

What do you think about this book? Or any other Arnold title?
Please send your comments to feedback.arnold@hodder.co.uk

*To Kathleen, who taught me the value of silence, exile and
cunning ...*

Contents

Acknowledgements

This is a work of synthesis and relies heavily on the writings and ideas of others, as the bibliography indicates. I have used and abused the knowledge and dedication of many librarians in Loughborough, Cambridge, London and Paris in pursuit of this material. It is a pleasure to acknowledge their assistance. The following pages also reflect conversations with friends and colleagues, notably the members of the Geography Department at Loughborough, past and present, who have provided an enduringly convivial environment in which to teach and research, despite the mounting pressures of academic life. A few have helped directly (though often unwittingly) in the preparation of this book, especially David Atkinson, Robin Butlin, Denis Cosgrove, Pyrs Gruffudd, Cheryl McEwan, Hayden Lorimer, Jason Roberts and Peter Taylor. Beyond my place of work, I must thank the editorial staff at Arnold who have shown great patience. Colleagues in other universities have provided advice, assistance and regular entertainment. Thanking them would be rather embarrassing – a despairing attempt to acquire credibility by association. I would like to offer, instead, my apologies to the following for wasting so much of their valuable time: Alan Baker, Mark Billinge, Louise Crewe, Stephen Daniels, Felix Driver, Derek Gregory, Gerry Kearns, Chris Philo, Graham Smith and Charles Withers. I've enjoyed their company immensely; I only hope they continue to endure mine.

Introduction: geography and the meaning of Europe

Anyone who speaks of Europe is wrong – a geographical notion. (Otto von Bismarck)[1]

The project of European economic and political integration is the main source of political controversy in all European Union countries and in many beyond. In Britain, where I live, two Conservative Prime Ministers have been undermined in the space of 10 years in the 1980s and 1990s by the bitter recriminations between 'Europhiles' and 'Eurosceptics'. These angry disputes have been accompanied by a growing 'Euro-mountain' of published words on every conceivable aspect of the 'European question' and in every imaginable language, from opinionated newspaper editorials to the most learned academic treatises. The intensity of the 'Euro-debate' shows no sign of waning. 'Europe', it would appear, is set to dominate the political process well into the twenty-first century.

This is scarcely surprising, for there are fundamental issues at stake here concerning the nature of the state, government, citizenship and sovereignty. These are visceral and emotive topics, the discussion of which tends to generate more confusion than enlightenment. While most people would agree that the old European nation-states can no longer function in their traditional manner in an increasingly globalised world, there is no consensus about the most appropriate alternative arrangement. As we nervously approach the end of the millennium, the sense of bewilderment and unease has become palpable. Europe's inhabitants (however they are defined) seem genuinely uncertain about where their ultimate allegiances lie and similarly confused about the most desirable scale at which government, sovereignty and citizenship should operate: the town or city, the region, the nation-state or some emerging European confederation? These are not mutually exclusive choices, of course, but they raise the most basic and divisive of human inquiries: who am I and where do I belong?

Many of us, it would seem, are most comfortable with the idea of the traditional nation-state. This is still the principal locus of most personal and collective loyalties. Our enduring faith in the 'old' nations is not because they are more 'real' or less problematic than any other form of abstract political and cultural organisation. As Benedict Anderson and Homi Bhabha remind us, nations are 'imagined communities' constructed through widely-accepted 'narratives' which hold together disparate constituencies in a common bond of belonging.[2] It is this imagined bond which allows some of us to cheer with an uncharacteristic abandonment in support of eleven soccer players whom we do not know sporting one kind of shirt during a game against eleven equally anonymous opponents wearing a differently coloured ensemble. This is a process of geopolitical invention, a ritual which allows us to assume, absurdly, that the eleven men we are supporting are performing in our name, on our behalf and at our behest.[3]

The idea of an imagined European community, which would allow us to cheer with comparable passion for a European soccer team, is far more difficult to envision. The symbols and anthems of 'Europeanism' are as yet unfamiliar and we lack a language, both literally and metaphorically, through which to articulate any common feeling of 'Europeanness'.[4] The very word 'Europe' is itself a stumbling block, for the term is used quite differently by rival political groupings. Those who are enthusiastic about a future European federation tend to see 'Europe' as a distinctive cultural arena united by a common historical experience and a shared civilisation. From their perspective, a united Europe would mark the resolution of a long historical process, the final sweeping away of an outdated political geography fashioned by the economic, social and cultural conditions of the eighteenth and nineteenth centuries. For those who are sceptical about (or even downright hostile to) European federation, the word 'Europe' is a short-hand for the institutions of the European Union, which are depicted as centralising, inflexible, corrupt and wasteful. From their viewpoint, European unity represents a bureaucratic nightmare, the imposition of a new European geopolitical order by a small, unrepresentative elite who seek to ride roughshod over the cherished complexities of national traditions and cultures. These terminological differences reflect substantive interpretational disagreements. For example, whereas European unity is presented by 'Euro-enthusiasts' as the only viable response to an increasingly globalised (post)modernity, 'Eurosceptics' dismiss the idea of a federal Europe as an unnecessary governmental tier, cutting across the exciting new connections between individual citizens and global processes. A united Europe is thus the logical outcome of 'globalisation' or an impediment to it, depending on your viewpoint.

This book is a modest attempt to consider the following question: what has Europe meant in the past and what might it mean in the future? This is by no means an unheralded inquiry, for several single-volume histories of the continent have grappled with this very question.[5] But these 'big-picture' surveys are concerned as much with events, processes and beliefs *in* Europe as with the idea *of* Europe. They can therefore be distinguished from a smaller body of literature which has engaged more directly with the changing meaning of Europe.[6] There are two related limitations of this literature which together open a space for a new, *geographical* consideration of the European idea.

First, many of these accounts are based on a 'prescriptive' view of Europe.[7] They assume that a united Europe is intrinsically desirable, the only viable alternative to the manifest dangers of a discredited nationalism. The history of the European idea is then read 'backwards' from the present into the past so that recent moves towards European unification appear as an inevitable historical evolution. Europe, in these accounts, appears as a 'self-positing spiritual entity that unfolds in history and never needs to be explained',[8] something to be celebrated rather than scrutinised. The inspirational becomes inherently European; the malign anti-European and usually nationalist. Good things are of Europe; bad things merely happen there. 'Prescriptive' views of Europe imply that entire historical periods are more or less 'European'. The same Mediterranean peninsula gives us the *European* Renaissance but *Italian* fascism.

The French historian Jean-Baptiste Duroselle once famously declared that 'between 1914 to 1945, there was no Europe'.[9] This period represented, he believed, the climax of 'anti-Europeanism', the era when nationalism and imperialism, transmogrified into fascism, destroyed Europe both materially and intellectually in a near continuous 'civil war'.[10] Professor Duroselle's writings were explicitly designed to avoid the parochialism of national histories and to celebrate the unity of Europe's past, present and future. His was a laudable, and perhaps a necessary, fiction. But it was a fiction nonetheless. Europe has always been a protean and much abused term and it is dangerous to assert that the concept is intrinsically associated with unity, peace and harmony. While Europe's wars are still recalled and commemorated by national governments (and hence give meaning to national rather than European identities), one cannot ignore the 'Europeanness' of these conflicts, as they were often waged and justified in Europe's name. Violence, I would suggest, is integral to the European collective experience, and to insist that warfare represents the negation of the European ideal is a kind of historical myopia, an invitation to accept one version of Europe at the expense of all others. Historically dubious in European terms, this becomes tragically ironic from an African or Asian perspective. Non-Europeans could reasonably claim that European unity has always existed, but only as a common impulse towards colonial expansion, domination and oppression.

The second limitation of the literature on the European idea is its resolutely *historical*, rather than *geographical*, character.[11] In most accounts, geography is regarded as ontologically insignificant, an immutable, unchanging and independent physical background to the swirling historical drama.[12] Trivialisation of the geographical is partly motivated by a desire to avoid the pitfalls of determinism, but it also reflects a deeper intellectual prejudice which emphasises the historical at the expense of the geographical, on the grounds that the former is the realm of ideas whereas the latter is the preserve of the uninterpreted physical fact or empirical detail. Virtually everything of interest becomes an 'historical' issue (including matters as intrinsically geographical as debates about the spatial configuration of states). The prosaic, the dull and, above all, the physical are deemed the proper concerns of geography. All the geographer can offer, it would seem, is factual information about river basins and mountain ranges, the political significance of which must be left to other, more sophisticated, commentators.[13]

Bismarck's quotation at the beginning of this chapter encapsulates this prejudice. On one level, his comment was entirely valid. After all, where was the European government or army in 1876? But it is the rejection of Europe as a *geographical* concept which is interesting here. For Bismarck, Europe existed only on a geographical level. It was, therefore, trivial or simply 'wrong', a cartographic fiction, an invented space on the atlas map. As such, it had no place in the language of politics or diplomacy.

The attempt to rescue the idea of Europe from Bismarck's dismissive interpretation has been mounted entirely in the name of history and at the expense of geography, the irrelevance of which is tacitly accepted. Europe, its defenders claim, is an historical rather than a geographical concept. Bismarck was wrong, therefore, because he ignored the existence of a transcendent and evolving European consciousness. Europe is no lifeless cartographic abstraction but a powerful historical project animated by the writings of prophets and activists, the founders of the modern programme of European integration.

The belief in Europe as a *meaningful* historical concept (as opposed to a *meaningless* geographical concept) reflects a wider Eurocentrism which has determined how Europeans have viewed the moral geography of the world over the past few centuries.[14] Until comparatively recently Europe has been depicted as the cradle of a 'hot', dynamic, civilisation rooted in the ancient past and still advancing into the future, constantly overcoming environmental, political and cultural impediments. Here, an unfolding, progressive history has given meaning to thoughts, beliefs and deeds. Europe can therefore only be understood historically. Inhabitants of 'less fortunate' regions, on the other hand, have usually been seen as 'cold', static cultures, as 'peoples without history', frozen in an unchanging, non-progressive time and embedded within their natural environment.[15] The 'underdevelopment' of Africa, for example, its 'failure' to ascend a ladder of progress invented by Europeans, has previously been attributed either to 'problematic' environments which Africans could not control or to racial failings. Before the arrival of European colonists, who brought with them a sense of dynamism and progress, African lives were imagined as if suspended in Malthusian cycles of overpopulation, famine and death. Time did not progress there; it simply repeated itself. There was no history in the European sense; only space, nature, geography. Africans were thus explicable in geographical rather than historical terms.[16] Such pervasive Eurocentrism has been reinforced in many ways. The familiar Mercator map of the terrestrial globe provides an obvious cartographic affirmation. Here, the land surface of the earth appears as if naturally grouped around a small European 'core'. Europe, *geographically* inconsequential compared with Asia, Africa or the Americas, demands to be seen as the *historic* centre of the world, the heart of all civilisation (Figure I).

In recent years, particularly since the publication of Edward Said's book *Orientalism* (1978), the nature of these Eurocentric assumptions have been seriously re-examined.[17] Said emphasised the constitutive significance of geography as a *discourse*, as a way of describing and analysing places and peoples. Geographical description, he claimed, was a central element in the 'invention' of the non-European world against which Europeans defined themselves. This insight has spawned an expanding literature on the idea of the Orient, Asia or Africa in the European 'geographical imagination'.[18]

This book develops this line of inquiry by offering a short intellectual history of the idea of *Europe* in the European geographical imagination. It examines how Europeans have creatively imagined themselves *geographically*. Consideration of Europe as a geographical idea subverts the intellectual structures through which the continent has conventionally been

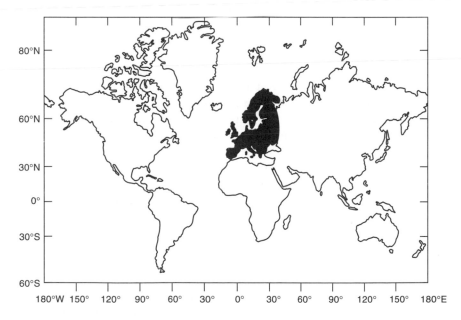

Figure I Europe and the European world-view: the Mercator projection.

interpreted. Geography is considered here not as a material force but as an intellectual arena of ideas and beliefs. Europe is likewise interpreted not as a singular historical evolution but as a contested geographical discourse; as a series of invented geographies which have changed over time and across space. What follows is, in this sense, an *historical geography* of the idea of Europe.[19] This is not to deny the enduring power of the material world, nor is it to undervalue the importance of the interaction between external nature and internal consciousness; it is, rather, to underline the importance of intellectual conceptions of geographical space in the formation of human consciousness.[20] As Henri Lefebvre has argued, space is not an independent given but a mutable and ever-changing product of economic, social, cultural and political processes.[21]

This book is by no means an exhaustive analysis (though it may be exhausting to read) nor does it present new evidence. It is based on texts from several disciplines but is aimed primarily (though not exclusively) at those studying the geography of modern Europe in universities and colleges of higher education. This is a big theme and I have sought to limit myself in two ways. First, the principal concern is with the idea of Europe as a *political* space, and this is reflected in the literature considered. Different accounts of Europe as an economic, social or cultural arena could (and should) be produced based on different literatures. Second, much of the book is concerned with the twentieth century, although Chapter 1 provides a much longer historical sweep. The focus on the past hundred years is governed by the needs of most geography students, for whom the distant past commands interest only insofar as it helps to explain the present. But very recent events also make a retrospective survey of the twentieth century conceptually appealing. Not all societies measure time in the same way, but in Europe the final years of centuries invariably produce anxious, *fin-de-siècle* speculation about the future in the light of the past. The decade since 1989, which marks the end not of a mere century but an entire millennium, has been no exception. It also began with an unexpected seismic shift in the European geopolitical order brought about by the fall of communism in central and eastern Europe and the subsequent collapse of the Soviet Union. The 'shock waves' of 1989–91 are still reverberating around the world and it will be many years before the implications of this period can be fully assessed. It is already clear, however, that the break-up of the Soviet empire has impacted most directly on Europe, where the political map of the eastern and central parts of the continent has changed beyond all recognition. The upheavals experienced since 1989 are rooted in the events which reconfigured the political geography of Europe between 1914 and 1945.[22] A focus on the twentieth century allows this most recent European revolution to be set in its appropriate historical and geographical context.[23]

This exercise raises an important methodological dilemma which it is important to confront at the outset. Given the objective is to destabilise the idea of Europe and to chart its different geographical meanings, it is obviously inappropriate to resolve these contradictions at the outset by advancing an *a priori* definition of what constitutes Europe. The purpose is to examine competing definitions on their own terms rather than in comparison with my own. Yet the absence of a working definition makes it difficult to identify a 'European' geographical imagination or a 'European' literature in which to trace these different visions. This becomes particularly significant in considering the writings of some British and Russian authors who commented on Europe while denying their own 'Europeanness'. In view of the obvious importance of British and Russian perspectives, such writings are considered alongside those of authors from France, Germany and Italy who rarely questioned their status as Europeans. North American views of Europe are also considered for the same reason: most twentieth-century images of Europe make reference to US perspectives, and the USA crucially influenced the debate on Europe after 1945. The important and still under-researched theme of 'non-European' images of Europe, as produced by Asians, Africans or Latin Americans, will not be considered.[24]

Notes to Introduction

1 This remark was scribbled into the margins of a letter Otto von Bismarck received from Prince Alexander Gorchavek, a Russian minister, in 1876. It is variously quoted and explained in Barraclough (1963, 33); Duroselle (1990, 333), Garton Ash (1993, 387 and 606) and Joll (1980, 10).
2 B. Anderson (1991); Bhabha (1990).
3 On geography and national identity, see Braudel (1986); Daniels (1993); Dijkink (1996) and Hooson (1994).
4 Fulbrook (1994).
5 It is equally central to H.A.L. Fisher's influential *A History of Europe*, published in 1936, and Norman Davies's magisterial *Europe: A History*, which appeared 60 years later.
6 Hay (1968); also Albrecht Carrié (1965); Barraclough (1963); Beloff (1957); Chabod (1965); (Duroselle 1966; 1990); Foerster (1967); Fuhrmann (1981); Geremek (1996); Gollwitzer (1951); Hale (1993); Heater (1992); Morin (1991); Nelson *et al.* (1992); Voyenne (1964) and Wilson and van den Dussen (1993).
7 The term *prescriptive* is stolen from Garton Ash (1993).
8 Delanty (1995, ix).
9 Duroselle (1966, 261).
10 The idea of a continuous European civil war is developed, for rather different reasons, by Nolte (1987).
11 J. Lévy (1997) is an exception. See C.A. Fisher (1966); Halecki (1950; 1952); McNeil (1974) and W.H. Parker (1960) on debates about the geographical extent of Europe, particularly in the east. Excellent essays on the role of geographical and environmental factors in shaping a European consciousness can be found in Braudel (1972; 1979); Pounds (1979; 1985) and C.T. Smith (1978).
12 Agnew (1989); Soja (1989, 10–42).
13 For a comparable recent critique which emphasises the need for a sociological rather than an historical reading of the European condition, see Mendras (1997).
14 Amin (1989); Davies (1996, 42–6); Lampropoulos (1993).
15 Wolf (1982).

16 The distinction between European 'historical geography' and North American 'cultural geography' can be interpreted in these terms. These remarkably similar intellectual activities are both concerned with the impact of past societies on the landscape. In Europe, the pioneers of *historical* geography saw themselves as chroniclers of an ancient European civilisation, an uninterrupted sequence in which past and present were seamlessly connected. In the Americas, although the evidence of ancient human occupation was scarcely less extensive, the past was conventionally divided into a pre-Columbian era, dominated by 'primitive' peoples, and a post-Columbian period of European expansion. 'History-as-progress' only began in the second era. The idea of an *historical* geography of the Americas only made sense in this later, European, phase. The study of the geographical legacy of 'a people without history' from the pre-Columbian phase demanded another disciplinary label. The late nineteenth-century invention of 'primitive culture' provided an alternative term – hence *cultural* geography.

17 See MacKenzie (1995) on the post-Said industry.

18 Said (1993); Godlewska and Smith (1994) contain up-to-date bibliographies. On the 'geographical imagination', see D. Gregory (1994).

19 For an 'historical geography of ideas', see Livingstone (1992). Recent historically informed work in political geography and 'critical geopolitics' informs this approach. See Dalby (1991); Hepple (1986); O'Sullivan (1986); Ó Tuathail (1996); Painter (1995); G. Parker (1985; 1988) and P.J. Taylor (1993).

20 Sack (1980).

21 Lefebvre (1991).

22 Halliday (1990, 11); Hobsbawm (1992a, 165).

23 A chronological structure is obviously somewhat problematic given my comments on the need to unsettle a complacent Eurocentric historicism which posits a single European idea evolving in an inevitable progression. Alternative structures are possible, of course, but they would be more challenging for the reader. I can only hope that the theme of Europe as a contested and imagined space emerges with sufficient force in each section to overcome any sense of historical inevitability.

24 Kiernan (1980).

1

Europe: the historical geography of an idea

Introduction

This chapter examines different visions of Europe from the Renaissance to the nineteenth century. It considers the origins and development of a distinctively European geopolitical imagination which placed increasing emphasis on territory as the primary basis of political identity and sovereignty.

Christian Europe: the origins of modern territoriality

How was Europe understood geopolitically at the dawn of the modern era? The political map of early sixteenth-century Europe, in the aftermath of Christopher Columbus's voyages, is a picture of geopolitical complexity, particularly in the German lands of central Europe where a multitude of small states and principalities existed in a kaleidoscope of overlapping allegiances and regimes (Figure 1.1). But even this chaotic image simplifies the historical reality. All maps are fictions, but a political map of premodern Europe in which states, territories and boundaries are described by clearly defined lines must be treated with particular caution. These lines are cartographic fictions, the backward projection of modern geopolitical conventions into an age where politics was structured by very different rules. Indeed, it is difficult to speak of 'Europe' at all in this period for the term had no currency in political debate.[1] Most accounts agree that, from the Classical world of Greece and Rome onwards, the Mediterranean (rather than the three surrounding continents of Asia, Africa and Europe) was seen as the central and most advanced of the world's regions, a pre-eminence explained by reference to environmental and climatic theories derived from Aristotle and Hippocrates.[2] The notion of Europe, or of any land continent, had little significance in classical or Medieval geography.[3] Significantly, the word 'Europe' does not appear in the Bible.[4]

A concern with global continental divisions emerged only in the post-Roman, Christian era, particularly from the seventh century AD when any residual sense of Mediterranean unity was shattered by the first great wave of Muslim expansion. The famous T-and-O *mappae mundi*, developed in the early seventh century by Isidore, Bishop of Seville, depict a circular

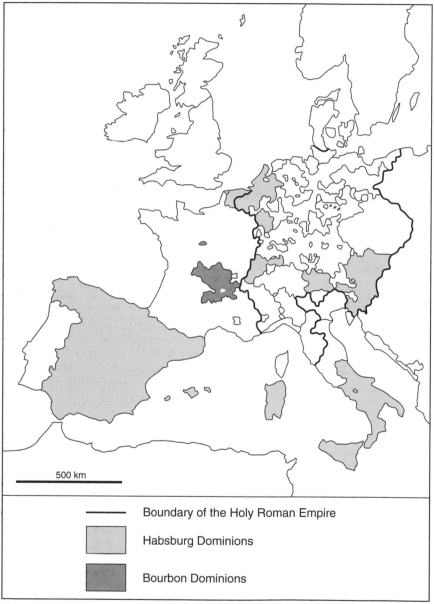

Figure 1.1 Europe in the wake of Columbus, *c.* 1520. Source: after C.H. Williams (1989, 198).

world divided into three unequal segments – Asia (representing half the earth) and Africa and Europe (representing one quarter each). These three zones were supposedly inhabited by the descendants of Shem, Ham and Japheth, respectively, the three sons of Noah.[5]

From this period, Islam and Christianity were on a collision course. Following the rise of Charlemagne, the emergence of the Holy Roman Empire and the launching of the first Crusades, an implicit association began to emerge between Christianity and the area which became Europe; between *Christianimus* (the faith) and *Christiantas* (the region).[6] By the twelfth century, maps devoted solely to the depiction of 'Europe', such as Lambert's *Liber Floridus*, were beginning to appear.[7]

Although the word 'Europe' was rarely used, the idea of 'Christendom' had become a meaningful geopolitical concept by the early fourteenth century.[8] Its relative unity was based on what Southgate calls the 'European superculture', derived from a common 'official' language of Latin and a general adherence to Christianity, both of which produced artefacts, symbols and practices (coins, statues, icons, legal charters and educational systems) which were widely understood and accepted.[9] These were the building blocks of what Denys Hay calls the 'continental ideology'.[10]

To be sure, the idea of a common, fourteenth-century European identity would have been limited to a small, educated elite who were in other respects divided by mutual suspicion and antagonism.[11] The fact that one of the first proposals for a united 'Europe' (devised by the French jurist Pierre Dubois as a means to defend Christianity against the Ottoman threat) dates from precisely this period is itself eloquent testimony that a sense of common 'Europeanness' was still relatively weak.[12] The cohesive power of Christianity in this period was, moreover, strictly limited because the Medieval church had been split between the Western Latin and Eastern Byzantine branches from the mid-eleventh century. It was only after the resurgence of Muslim power, culminating with the fall of Constantinople to Mehmed II in 1453 and the ensuing collapse of the Byzantine church, that a more culturally and religiously homogeneous sense of a western, Christian Europe emerged. By the late fifteenth century, the idea of *Respublica christiana* was frequently evoked in Papal proclamations with specific reference to the ominous external threat posed by an Islamic 'Other' into whose covetous hands fell the last vestige of Christian Greece (Morea) in 1460. The 'enemy' was now at the gates and the Christian realm was on the defensive. In 1521 Belgrade succumbed and by the end of that decade Vienna itself was under siege.

A more unified Christendom, though it was never without its internal divisions and disputes, developed in the midst of a terrible and seemingly irresistible external threat. These changes can be discerned in the world maps of the early Renaissance in which 'Europe' was usually depicted as a small and exclusively Christian region.[13] Hesitant though it was, the emergence of Latin Christendom engendered the first stirrings of a new political territoriality, a Christian geopolitics in which the spatial extent of Christendom assumed new significance. This was revealed, *inter alia*, by the angry fourteenth-century and early fifteenth-century debate about the correct location of the Holy See, a dispute sparked off by Pope Clement V's decision to transfer his residence from Rome to Avignon. This was subsequently justified on the grounds that the French city was closer to the geographical centre of the Christian world, an argument which clearly implies a spatial vision of Christendom.

However, the growth of Latin Christendom in the early centuries of this millennium (in the face of fierce Muslim resistance) demonstrates that the Medieval idea of political space was still radically different from the modern concept. This theme has been explored by Robert Bartlett in his work on Christian expansion from the mid-tenth to the mid-thirteenth century, an era in which Christendom virtually doubled in size. Bartlett concludes that the territorial growth of 'Catholic Europe' did not lead to the kind of regional distinctions between core and periphery which have been such a feature of later forms of imperial conquest. He claimed:

> The net result of this [Medieval] colonialism was not the creation of 'colonies', in the sense of dependencies, but the spread, by a kind of cellular multiplication, of the cultural and social forms found in the Latin Christian core. The new lands were closely integrated with the old. Travellers in the Middle Ages going from Magdeburg to Berlin and on to Wroclow, or from Bruges to Toledo and on to Seville, would not be aware of crossing any decisive social or cultural frontier.[14]

Driven by consortia of Latin priests, Frankish knights, merchants and travellers, '[h]igh medieval colonisation was . . . a process of replication, not differentiation'. The 'borders' between the Christian and Muslim worlds were also, it seems, transitory and far from clear-cut.[15]

This should not be overstated, of course, for the conflict between Christian and Muslim was self-evidently territorial in some degree. However, the Medieval rivalry between Christian and Muslim was concerned more with sites or cities of religious or symbolic significance rather than with land as a strategic or commercial possession. The struggle for control of Jerusalem and the Holy Land, beginning with the first European Crusades, is the most obvious example, though the attempts by late fifteenth-century cartographers such as Christopher Buondelmonti to persuade his Renaissance audience that Muslim Constantinople was really part of Christendom reveal a comparable level of proprietorial anxiety with regard to that city.[16]

There are a number of reasons for the weakly developed sense of early Christian territoriality. Economically, the idea of private property was less than fully developed under the prevailing feudal system. As a result, land and territory were not sharply demarcated.[17] Politically, the Medieval order was based on multiple loyalties and complex allegiances operating in an overlapping and essentially aspatial fashion.[18] The idea of compartmentalised political space was, therefore, somewhat alien to the Medieval Christian world-view, and the familiar modern notion of a hierarchical spatial order, of cores and peripheries, was less evident. The essential characteristic of Medieval Europe's external limits and internal organisation was a relatively undifferentiated economic and political space, lacking clear, identifiable borders and frontiers.

The spatial fluidity of Medieval geopolitics was by no means unique to the Christian world. It was also discernible in Islamic culture where even the idea of 'area', familiar enough in Arabo-Islamic mathematics, had not developed as a geographical concept.[19] Arguably, the reluctance to conceptualise space as a commodity to be mapped and delineated persisted in the non-European world well into the modern era when, as we shall see in the next section, a very different European understanding of political space began to emerge.[20]

Secular Europe: the balance of power

The association between Christianity and Europe remained strong throughout the early modern period. In 1625, Samuel Purchas argued that Christ himself reflected a quintessentially European spirit:

> Europe is taught the way to scale heaven not by mathematical principle, but by divine verity. Jesus Christ is their way, their life; who hath been given a bill of divorce to ungrateful Asia, where he was born, and Africa, the place of his flight and refuge, and is become wholly and only European. For little do we find of his name in Asia, less in Africa, and nothing at all in America, but later European gleanings.[21]

Early modern ideas of Europe were, in this respect, an extension of earlier, Medieval, ideas of Christendom.

But the momentous and interwoven economic and intellectual changes associated with the rise of European capitalism and the expansion of European influence into the 'New World' gradually transformed and secularised the idea of Europe. Why capitalism and its associated geopolitical transformations took root in Europe is a widely debated point. While many commentators accept that Europe must have been exceptional, there is no consensus about the nature of this peculiarity. E.L. Jones stresses the physical geographical and environmental advantages which Europe possessed in comparison with other regions, including those (such as China) which had more advanced economic and technological systems at the end of the Medieval period.[22] Michael Mann emphasises European intellectual developments in the early modern period, particularly the emergence of what he calls, after Max Weber, Europe's 'rational restlessness'. This, combined with a post-Reformation Christian morality, laid the foundations for early European capitalism and its associated modern political systems and ideologies. It was a concatenation of new intellectual and cultural values which set Europe apart from the rest of the world and created its uniquely expansive outlook.[23] Jim Blaut entirely reverses this accepted logic, which he dismisses as 'Eurocentric'. Fifteenth-century Europe was in no sense exceptional, either actually or potentially. The advantages which it gained flowed from the 'fortunate' accident that the Atlantic is narrower and easier to traverse than is the Pacific. Europeans therefore arrived in the Americas before the peoples of Asia. The Atlantic trading system was thus established centuries before the Pacific Rim emerged as an oceanic economic space. It was the unrestrained European exploitation of the 'New World' which fuelled the emergence of European capitalism rather than the other way around.[24]

Whatever the causes of European capitalism and expansion after *c.* 1500, the consequences of these related developments were profound and far-reaching. Europe's increasing engagement with, and knowledge of, the distant regions of the globe allowed educated Europeans to develop an entirely new world-view and a new assessment of their relative position within the global scheme. It could be argued that the modern concept of geographical – and hence geopolitical – space was first invented in this period. The shift from a premodern to an early modern consciousness seems to have been accompanied by a movement from vertical (religious) conceptions of sacred space to horizontal (secular) notions of geographical space.[25]

The invention of modern political space in this period has been expertly examined by several authors.[26] We have already noted how post-Renaissance world maps, particularly those based on the Mercator projection, reflected a European perspective on the world. This was no longer a Medieval, Christian view of the globe as viewed from Jerusalem or Rome. Here was the earth's terrestrial space as seen from the brash new maritime cities of northern Europe and the Atlantic – from Lisbon, London or Hamburg. This was a commercial, capitalist vision in which Mammon had triumphed over earlier forms of religiosity. The old idea of Christendom gave way to the new, secular, notion of 'Europe', a word derived from the Greek *Europa*.[27] Europe, paradoxically appearing smaller and smaller as cartographic conventions and global horizons expanded, was depicted in ever-increasing detail, notably in the *Theatrum Orbis Terranum* (1570), the world's first modern atlas drawn up by Mercator's great rival, Abraham Ortelius (Figure 1.2).

There remained, to be sure, a tradition of depicting Europe in a more stylised, symbolic form, particularly as a human body. The most striking example is the 'map' of Europe as a Queen, reproduced in 1588 by Sebastien Münster in his *Cosmographia Universalis* (Figure 1.3). This image, designed as Habsburg propaganda, depicts Spain as the Queen's head and crown, with Bohemia as her heart. Gallia and Germania (France and Germany) are emblazoned across the Queen's chest, as the right and left breasts. The figure's right arm represents Italy, with the orb of Sicily held in

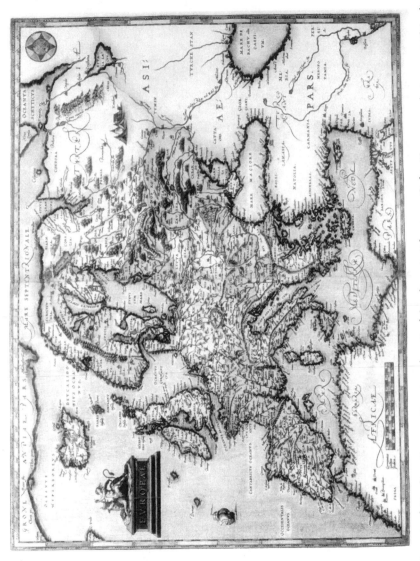

Figure 1.2 The Europe of Ortelius. Source: *Europa*, plate 2 of Ortelius, A. *Theatrum Orbis Terrarum*, Antwerp, 1584 edition. Note: see Jacob (1992, plate 7, 112–13); Karrow (1993)

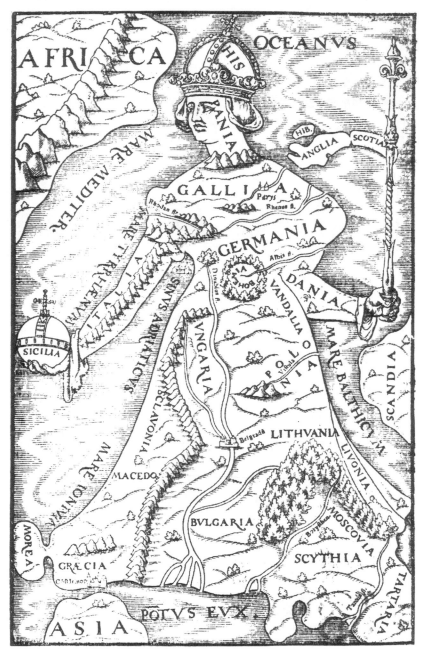

Figure 1.3 Europe embodied: Sebastien Münster's Europe. Source: folio xli in
Münster, S. *Cosmographia Universalis*, Basle, 1588. Note: this image has been
widely reproduced; for example, see Agnew and Corbridge (1995, 53); der Boer
(1993, 52); Davies (1996, xviii); Hay (1968, frontispiece) and Wintle (1996, 81).
General discussions are provided in Wintle (1996; 1999).

the outstretched hand. Other regions of central and eastern Europe are scattered about the flowing gown, the folds of which represent the continent's mountain ranges and river valleys. The only cities indicated are Paris, Belgrade and Constantinople. The Queen's flexed left arm represents Denmark, and in her left hand is a sceptre with a detached and rather tattered pennant representing England, Scotland and Ireland. There are few more eloquent statements of the marginal position of the British Isles in the Habsburg view of sixteenth-century Europe.

The secularisation of the European self-image was influenced by the breakdown in the fragile unity of the Christian church following the Reformation and the emergence of thriving Christian communities (Catholic and Protestant) in other continents. The idea that Europe was the sole arena of a united Christian faith was no longer tenable. But the emergence of a modern system of European nation-states substantially altered the meaning of political space, both in Europe and across the globe as European power expanded.[28] Before the sixteenth century, the word *state* referred simply to the status of the ruler and his or her domain and had few territorial implications. In the course of the next century or so, the idea of the state as a geographical arena slowly gathered momentum. The *Six Livres*, written by the French jurist Jean Bodin in 1576, was of considerable importance in establishing the concept of a spatial, centralised state.[29]

The 'territorialisation' of power during the seventeenth century became manifest in several ways. As Chandra Mukerji shows in her study of the gardens at Versailles, the shaping and representation of geographical space, including landscape design, became a central component of the exercise of absolutist state power.[30] But it is the visual representations of Europe and its constituent territorial states, particularly the cartographic images, which again betray this shift most clearly.[31] Early modern maps were profoundly political documents – carefully constructed self-portraits of the new European nation-states designed to legitimate their territorial authority.[32] As I have already suggested, political frontiers and borders were previously understood only in the vaguest terms as zones of transition rather than as sharp lines of demarcation. Rarely were such divisions expressed cartographically on international agreements before the eighteenth century.[33] Gradually, however, national frontiers were surveyed, mapped and then policed on the ground. The great Cassini survey of France, inaugurated in the late seventeenth century by Jean-Baptiste Colbert and endlessly updated through the eighteenth century, was simply the most ambitious example of a broader development: nation-building through mapping.[34]

This 'fixing' of national spaces on the European map took place in the midst of widespread violence and warfare, particularly during the Reformation, the Counter-Reformation and the Thirty Years War, a conflict which claimed up to 30 per cent of the population in parts of central Europe. These traumas engendered a new international geopolitics best exemplified by the idea of the balance of power which has dominated

international relations and the European interstate system since the sixteenth century.[35] One of the earliest prophets of this concept was Niccolò
Machiavelli, an official in the Florentine chancellery whose essay on *The
Prince*, written in 1513, first laid down the principles which were eventually to govern modern statecraft and international relations.[36] Although
primarily concerned with the labyrinthine political intrigues of Italian city-
states, Machiavelli's ideas were clearly relevant to the political system of
Europe as a whole.[37] His vision of international relations was based on the
metaphor of a set of scales. The ideal situation was a balance of power
between the rival dynasties – on the one hand the French Valois, and on
the other the Habsburgs, who dominated Spain, most of Italy, the Low
Countries, Austria, the German lands and Burgundy. Relatively independent nation-states, notably England, were of critical importance, for they
held the balance of power. This finely tuned system of checks and balances
was the best, perhaps the only way to achieve peace and stability in a
Europe where clashes between dynastic houses and tensions between religious faiths threatened to wreak unprecedented havoc.

The balance of power was initially developed as a strategy to keep
Habsburg power in check. Following the collapse of the Spanish Armada in
1588 and the general diminution of Habsburg influence during the seventeenth century, the same notion was then used to challenge the expansive
ambitions of France, especially under Louis XIV.[38] Increasingly, the idea of
Europe came to be defined by France's mainly Protestant northern
European enemies in a somewhat utopian fashion as the political arena in
which a balance of power between rival nation-states might conceivably
operate and in which war and destruction might ultimately become impossible. Thus was born the modern idea that the concept of Europe is somehow naturally associated with the cause of peace and harmony; an article of
faith rather than a statement of observable geopolitical reality.

The defining moment in the development of the new European geopolitics was the Treaty of Westphalia in 1648 which brought an end to the
Thirty Years War and established the idea of international law along lines
laid down by the Dutch jurist Hugo Grotius.[39] The Treaty of Westphalia
was, according to Peter Taylor, 'the foundation treaty for the establishment
of the inter-state system'; the rock upon which the secular idea of Europe as
an arena of separate nation-states held in a balance of power was constructed (Figure 1.4).[40] Threats to the geopolitical balance were depicted as
threats to the very idea of Europe itself. When Louis XIV's armies attacked
the Dutch United Provinces in 1672, pamphlets streamed from Europe's
printing presses denouncing France's disruption of the balance of power, the
essence of what it meant to be European. From then until 1713, France was
virtually permanently at war around its eastern frontiers, leading to a significant expansion of French territory (Figure 1.5). French propaganda
spoke of the Sun King as the defender of the older idea of *Respublica christiana*. English Whigs and the Dutch Royal House under William of Orange,

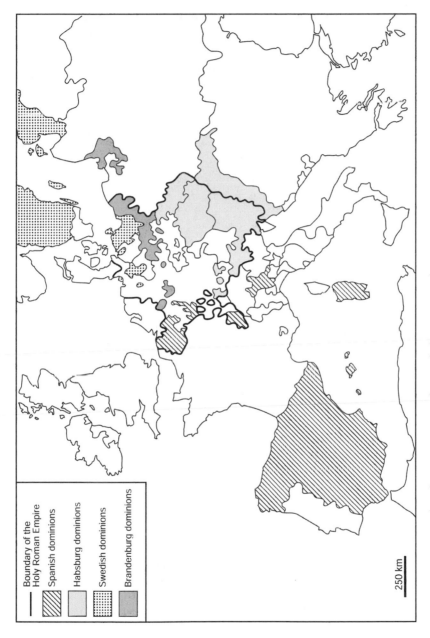

Figure 1.4 Europe after the 1648 Treaty of Westphalia. Source: after Merriman (1996, 175).

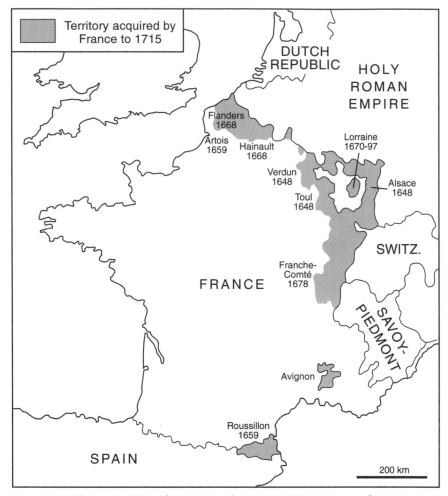

Figure 1.5 The expansion of France under Louis XIV. Source: after Merriman (1996, 318).

united constitutionally following the Glorious Revolution of 1688, saw themselves as struggling to preserve the religious freedom and commercial liberty of Europe against the intolerance of an older, Catholic geopolitical order.[41] It was 'in the course of the seventeenth and early eighteenth centuries', claims Denys Hay, that 'Christendom slowly entered the limbo of archaic words and Europe emerged as the unchallenged symbol of the largest human loyalty'.[42] Schmidt goes further and offers what he calls 'conclusive evidence that the term Europe established itself as [an] expression of supreme loyalty in the fight against Louis XIV. It was associated with the concept of a balanced system of sovereign states, religious tolerance, and expanding commerce'.[43] As a political expression, Europe emerged as a secular, if not a Protestant, idea of political liberty which could be preserved by a just balance between rival nation-states. Although increasingly deployed through the sixteenth and seventeenth centuries, it only found real currency and influence after *c.* 1700.[44] Significantly, the Treaty of Utrecht, signed in 1713 at the end of the War of Spanish Succession, is the last European accord to make reference to *Respublica christiana*.[45]

The most prophetic vision of Europe within the early modern secular geopolitical imagination was formulated by the Duc de Sully (Maximilien de Béthune), Henri IV's shrewd finance minister.[46] After the assassination of Henri in 1610, Sully retired from public life and, supported by Cardinal Richelieu, developed a 'Grand Design' for European unity and peace which appeared in several forms during the middle decades of the seventeenth century, usually attributed to his former royal patron.[47] Whether Sully was genuinely concerned to promote European reconciliation or was motivated instead by a desire to diminish the power of the Habsburgs and reflect well on the ambitions of the Bourbons is a matter of serious debate. His legacy has been lasting, however, not least because of his realistic acceptance of geopolitical realities and his advocacy of peace both on moral and on political grounds. Sully was no utopian dreamer – he even described himself as 'cold, cautious and unenterprising'.[48] If states were no longer influenced by religion, he reasoned, they may yet be persuaded by political economy. The ultimate objective was 'to divide Europe equally among a certain number of powers and in such a manner that none of them might have cause either to envy or fear from the possessions or power of others'.[49] Once Habsburg power was limited to its Iberian heartland (together with its non-European colonial possessions), Sully envisioned 15 of the existing nation-states combining to form a new European order. Six of these were hereditary monarchies (France, Spain, England, Denmark, Sweden and Lombardy), five were elective monarchies (the Holy Roman Empire, the Papacy, Poland, Hungary and Bohemia) and four were republics (the Venetian, the Italian Ducal, the Swiss and the Belgian). Each state would provide proportional quantities of men and resources to establish a common European army, obviating thereby the need for rival national armies. A European general council, as well as several regional sub-councils, were also envisaged, the former to be located 'in the centre of Europe' on the borders of Frankish and Germanic lands. Metz, Luxembourg, Nancy, Cologne, Frankfurt, Heidelberg, Speyer, Strasbourg, Besançon, Basle and Trier were all suggested as possible seats of the general council (Figure 1.6). Sully's description of England is hauntingly familiar to modern readers. The English, he observed, believed that

> the Britannic isles ... had never experienced any great disappointments or misfortunes, but when their sovereigns had meddled in affairs out of their little continent. It seems, indeed, as if they were concentred in it even by nature, and their happiness appears to depend entirely on themselves and their having no concerns with their neighbours.[50]

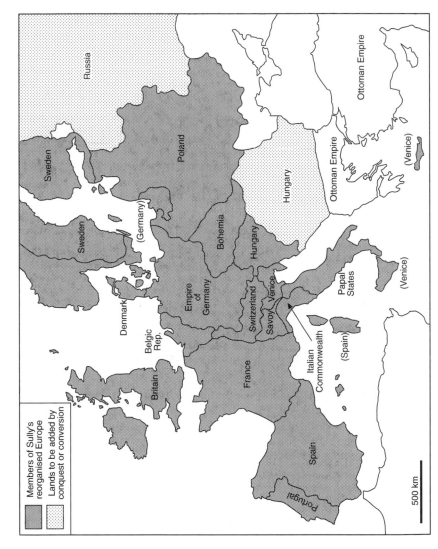

Figure 1.6 Sully's Grand Design. Source: after Heater (1992, 29).

It would be wrong to interpret Sully's project in entirely secular terms. As a Protestant he was well aware of the divisive potential of religious beliefs. Partly for this reason, he insisted on the need to preserve the existing status quo between Catholic and Protestant Europe. He also specifically rejected the possibility that non-Christian nationalities could be classed as part of Europe. In Sully's world-view the Ottoman Empire remained a principal 'Other' against which Europeans could seek common cause.[51] In one crucial respect, however, Sully's analysis was based on a form of reasoning which owed nothing to the traditions of Christian geopolitics. Russia, Sully insisted, had no place in the European order. The exclusion of Russia was less to do with its religious distinctiveness (it was, after all, a Christian place) than with Sully's firm belief that the Russian people were culturally inferior:

> I say nothing of Muscovy and Russia . . . [the people of] these vast countries . . . being in part still idolators, and in part schismatics, such as Greeks and Armenians, have introduced so many superstitious practices in their worship, that there scarce remains any conformity with us among them; besides, they belong to Asia at least as much as to Europe. We may indeed almost consider them as a barbarous country, and place them in the same class as Turkey, though for these five hundred years, we have ranked them among the christian powers.[52]

Sully's exclusion of Russia on cultural (rather than specifically religious) grounds had considerable significance for eighteenth-century conceptions of Europe's political space.

Europe and its Others: civilisational geopolitics

During the eighteenth century the national frontiers of the European map became more clearly inscribed in the geographical imagination of the continent's peoples. Europe's central position on the world map was reinforced by the addition of yet another 'new' continental land mass, Australasia, which appeared suddenly on new maps as if brought into being by the exploratory heroics of European navigators and seamen.[53] At the same time, the intellectual upheavals of the Enlightenment significantly altered the debate about Europe's characteristics, extent and internal geopolitical order.[54] Of critical importance was the emergence of a cultural (what Agnew and Corbridge call a 'civilizational') geopolitics.[55] This was prefigured by Sully's blueprint for European union but developed strongly in the eighteenth century to supplement, and partially to replace, existing ideas about the nature and distinctiveness of Europe. This is not to suggest that the idea of Europe as the privileged haven of liberty, tolerance, reason, science and industry was less important after 1700. However, it was only in the eighteenth century that secular theories of the human condition and of Europe's place within the global order were fully developed. Early examples were the climatic and environmental theories put forward by Charles Louis de Secondat, the Baron de Montesquieu, to explain Europe's apparent moral and political distinctiveness from the 'despotic' Orient, most famously in his *De l'esprit des lois* (1748).[56] Whereas the mild climates and varied environments of Europe created order and balance in human relations, in the more extreme conditions of Asia, where hot and cold countries coexisted without natural barriers, it was possible for strong, energetic peoples to conquer and dominate weaker and less energetic peoples by force of arms. In Europe, states were relatively small in size and thus more conducive to the rule of law and just government; in Asia, states were large and controllable only by force. Asia's climate was thus the breeding ground of 'Oriental despotism'.[57] But Montesquieu's vision, like that of Edward Gibbon, was still dominated by an older, circular concept of time and human history. Modern Europe had emerged from the ashes of the Roman Empire and was, he surmised, destined to be replaced by a new and superior order elsewhere, probably rising in the Americas. This somewhat pessimistic view was a central theme in William Robertson's *View of the State of Europe* (1769).

But there were other, more optimistic, views of Europe's unquestionable dominance. After 1750 a new lexicon of words and concepts, hitherto unknown or differently used, began to appear in moral and political debate and in the newly constructed dictionaries and encyclopaedias. Three of the most important expressions were the interrelated terms 'culture', 'civilisation' and 'progress', a secular trinity of Enlightenment thought which featured with increasing regularity in discussions of Europe's nature and distinctiveness. 'Culture' and 'civilisation', despite subtle differences in French, German and English, were used from the 1760s to indicate the peculiarly European characteristics of dynamism, energy and development mixed with logic, reason and self-restraint. These qualities came to be seen as the principal causes of European superiority over the more 'primitive' peoples of the Americas, Africa and Asia.[58] By 1766, in the midst of the Pacific explorations, the term 'civilisation' (first coined a decade earlier by the physiocratic thinker, the Marquis de Mirabeau) was twinned with its familiar adjective *européenne* in a treatise on French colonies in North America by the Abbé de Baudeau to contrast Europe, at the highest stage on the ladder of human progress, with the Americas, at the most primitive and savage. The same word and concept then appeared in English, notably in Adam Smith's famous economic treatise on *The Wealth of Nations* (1776).[59] The eighteenth-century notion of 'civilisation' was thus closely connected to the debates about the nature and potential of human progress which took place around the same time, particularly the materialist theories of human development conceived as the endless march of civilisation onward and upward through a series of stages from barbarism to enlightenment.[60]

This cultural rhetoric began to influence definitions of Europe's geographical limits, its internal geopolitical identity and the very nature of government and statecraft. This was connected to a more general Enlightenment shift in the language of politics. It is now widely accepted that modern notions of collective, national political identity emerged in Europe during the eighteenth century when the complex array of national icons, symbols and cultural practices first began to influence the values and outlooks of the great mass of European peoples.[61] Eighteenth-century nation-states increasingly defined themselves by reference to the number, wealth and contentedness of their subjects or citizens. New and more subtle political and constitutional relationships developed which amounted, according to Michel Foucault, to a fundamental change in the character of government and state power, a change associated with the invention of 'population' as a new object of inquiry.[62] Brute force and violence were no longer the principal forms of governmental authority (although they remained the ultimate sanction); power henceforth resided in a more complex dialectic between those who governed and those who were governed, between rulers and ruled.[63] This relationship was based on a degree of mutual consent, a collective consensus or common will, what Jean-Jacques Rousseau termed *la volonté générale*.[64] Modern nation-states were, therefore, far more concerned than their predecessors with the people that lived within their borders. This should not be seen as a benign concern on the part of the ruling elite for the welfare and happiness of ordinary folk. Rather, the social and political contract which developed between the state and the citizen (even before the emergence of participatory democracy) was part of a broader reconfiguration of European political culture and a realisation that people were both the form and the instrument of government, the ultimate source of national power and thus the foundation stone upon which most Enlightenment utopias, such as David Hume's idea of the 'perfect commonwealth', were constructed.[65] Nation-states would henceforth be identified by reference to the number and activities of their populations rather than the personalities of their rulers. The new language of political economy, often expressed in abstract numerical or statistical form, perfectly reflects this shift.

Debate about the geopolitical character and extent of Europe reflected these changes. The earlier preoccupation with the relationships between kings, princes and governments gradually shifted towards a cultural, even a quasi-anthropological, concern with the different peoples of Europe, now seen as the most important indicators of the continent's internal order and external limits. The concept of the balance of power, and its conceptual link with 'Europeanness', remained firmly in place; indeed the eighteenth century is usually seen as the 'golden age' of this idea.[66] However, Enlightenment geopolitical debate was increasingly articulated in a relatively novel cultural register. At the same time, the relationships between European states were becoming evermore complex. Britain, increasingly conscious of its extra-European imperial status, still maintained the balance between Bourbon and Habsburg, but this relatively simple arrangement was disturbed by the emerging eastern powers of Prussia under Frederick the

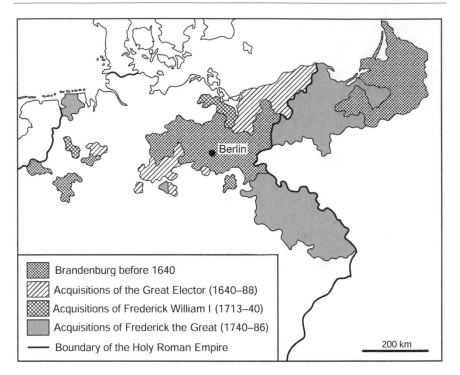

Figure 1.7 The rise of Prussia. Source: after Merriman (1996, 304); Lee (1984, 194).

Great (Figure 1.7), Russia under first Peter the Great and then Catherine the Great (Figure 1.8) and the associated collapse of Poland (Figure 1.9). Gradually, a more complex version of the European balance of power emerged based on a 'pentarchy' of five major powers – Britain, France, Austria, Prussia and Russia – hesitantly inaugurated by the Treaty of Utrecht (1713) (Figure 1.10).[67]

This new geopolitical order raised fundamental spatial questions which have bedevilled the debate about the nature and extent of Europe ever since. As Jeremy Black has shown in his sweeping historical account, the European credentials of 'northern' powers such as Britain were the subject of some debate, both within Britain and elsewhere on the continent.[68] Most pressing of all, however, was the perennial dispute about Europe's eastern limits. As we have seen, Sully drew a clear line between Russia and the rest of Europe in the mid-1600s based on cultural or civilisational criteria. By the early eighteenth century this division seemed increasingly problematic as Peter the Great's Russia was playing a more important role in European affairs. Peter accepted the inherent superiority of Western culture and civilisation and sought to change the face of Russia from a backward Slavic, Asiatic power to a modern, enlightened European nation, whilst simultaneously expand-

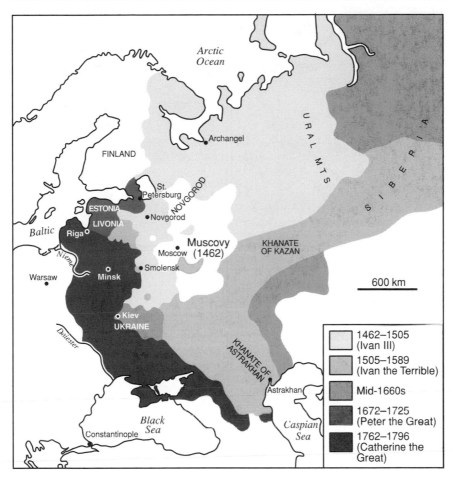

Figure 1.8 The rise of Muscovy/Russia. Source: after Merriman (1996, 315); M. Gilbert (1972, 41).

ing its territorial extent through almost constant warfare and settlement.[69] This was to be achieved literally as well as metaphorically, for Peter famously banned the wearing of beards throughout his empire in an attempt to undermine the power of Orthodox, and luxuriously bearded, clerics. As Mark Bassin has shown, this required energetic geopolitical propaganda to assert Russia's status as a European power which, like Britain and France, possessed a non-European empire.[70] The only difference was that Russia's extra-European empire was contiguous with the European core. The critical question, therefore, was where best to locate the dividing line between European Russia and its Asian empire; a border which Peter believed would mark the eastern limits not only of Russia but of Europe as a whole.[71] Older

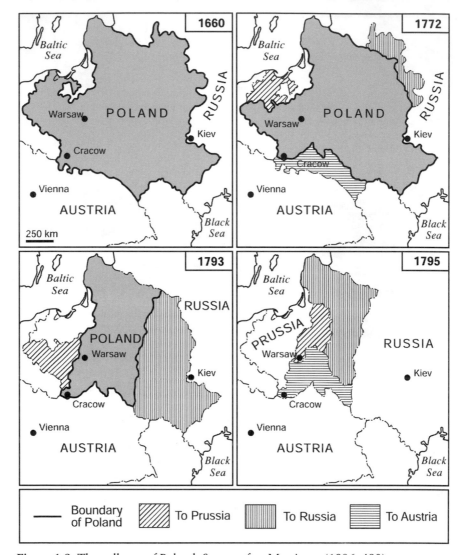

Figure 1.9 The collapse of Poland. Source after Merriman (1996, 490).

ideas that Europe and Asia were divided along fluvial lines, notably by the River Don, were quickly rejected in favour of the north–south spine of the Urals. This clear and immobile barrier became European Russia's official line of cultural and political demarcation and was mapped and remapped at a variety of scales throughout the eighteenth century.[72] Boundary posts were eventually erected to mark the eastern 'frontier' along the trail from Yekaterinburg to Tyumen and it was here that the Tsar's Siberia-bound prisoners would pause to scoop a last handful of European soil before entering their Asian exile.[73] When Catherine the Great ascended to the throne she

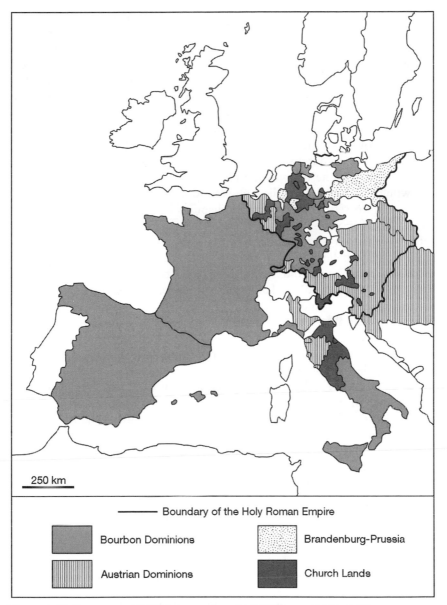

Figure 1.10 Europe, *c.* 1740. Source: after C.H. Williams (1989, 202).

was able confidently to assert (in French, of course) that 'La Russie est une puissance européenne'.[74]

Peter's ambitions to forge a new, European identity for Russia were endorsed by some commentators in the west. During the crisis of the late seventeenth and early eighteenth century several new projects for European peace and unity directly addressed Russia's European credentials and came to conclusions quite different from those of Sully. The English Quakers William Penn and John Bellers offered intriguing views on how to eradicate war from the face of Europe. Penn, scion of a wealthy aristocratic family who embraced non-conformism as a student at Oxford, fled to America in 1682 to avoid persecution. Here he established a Quaker colony in the state which still bears his name, Pennsylvania.[75] In 1693 he wrote a passionate, 8000-word essay on the 'groaning state of Europe' and the urgent need for a European Diet or parliament to remove the possibility of war.[76] Penn's abhorrence of war is manifest throughout this extraordinary document: 'He must not be a man but a statue of brass or stone', he begins 'whose bowels do not melt when he beholds the blood tragedies of war'.[77] His proposed Diet would meet in a symbolic circular room somewhere in Europe at least once a year to settle grievances and disputes by negotiation and majority voting. Its composition would be based on the scientific calculation of each state's population and productivity (an early estimate of gross domestic product). Penn believed this would produce a council comprising ten representatives from France and Spain, eight from Italy, six from England, four each from 'Sweedland', Poland and the Netherlands, three each from Portugal, Denmark and Venice, two from Switzerland and one from the Dukedoms of Holstein and Courland. Both the 'Turks and Muscovites', so firmly excluded by Sully, were welcomed by Penn 'as seems but fit and just'. Each would be allocated 10 places in the proposed Diet.[78] All states would be responsible for nominating their own representatives, and the language of debate would be either French ('very well for civilians') or Latin ('most easy for men of quality').[79]

The rest of the document counters the fears Penn imagined such an arrangement would provoke. The end of war in Europe would not 'endanger an effeminacy by such a disuse of soldiery', he insisted, 'because each sovereignty may introduce as temperate or severe a discipline in the education of youth as they please'. A common European army would still be able to resist external aggression. In words which echo down the centuries, he also claimed the Diet would not compromise the domestic sovereignty of nation-states, except insofar as military power was concerned: 'And if this be called a lessening of their power, it must be only because the great fish can no longer eat up the little ones, and that each sovereignty is equally defended from injuries'.[80] Thus would war come to an end:

> by the same rules of justice and prudence by which parents and masters govern their families, and magistrates their cities, and estates their republics, and princes and kings their principalities and kingdoms, Europe may obtain and preserve peace among her sovereignties.[81]

Bellers developed his own far-reaching version of Penn's scheme 17 years later.[82] Bellers saw the 1707 Act of Union between England and Scotland as a model for a future Europe. He advocated accepting whatever geopolitical order emerged from the ongoing War of Spanish Succession. A European

Parliament should then be established to redraw the political map of Europe. The objective should be '100 Equal Cantons or Provinces, or so many, that every Sovereign Prince and State may send one Member to the Senate at least'.[83] This arrangement would need to evolve gradually but, once in place, each canton would be required to contribute men and resources to a common European army. Only through this rational reorganisation of Europe's political geography could war be avoided in the future: 'For nothing makes Nations, and People more Barbarous than War; so Peace must be the first step, to fit Mankind for Religion'.[84] Like Penn, Bellers insisted that Russia and the Ottoman Empire should be welcomed into a reordered European state.

Two years later, in the midst of the negotiations for the Treaty of Utrecht, another Frenchmen, the Abbé de Saint Pierre, produced a hugely expanded version of Sully's Grand Design.[85] Saint Pierre's well-meaning but pedantic style was cruelly mocked by his contemporaries and the verdict of posterity has been equally uncharitable: one authority calls him 'the great bore of eighteenth-century France'.[86] Like Sully, Saint-Pierre's vision was compatible with the balance of power but, in common with earlier formulations, he argued that balance should be maintained by parliamentary dialogue rather than periodic warfare, an ideal enshrined in Grotius's *De jura belli et pacis*.[87] Like his English Quaker contemporaries, Saint-Pierre welcomed Russia into the European fold and was also more sympathetic towards the Muslim world than was Sully. The response to Saint-Pierre's writings was distinctly underwhelming. Frederick the Great wrote sarcastically to Voltaire that 'The Abbé de Saint-Pierre has sent me a fine work to re-establish peace in Europe. The thing is very practicable: all it lacks to be successful is the consent of all Europe and a few other small details'.[88] The witty Sainte-Beuve had fun describing the Saint-Pierre verbosity:

> He expands a sound idea, he proposes a useful reform. You approve it, he is not content; in order to establish it more firmly, he goes on to amuse himself by listing the most futile objections, giving himself the pleasure of refuting them one by one; firstly, secondly ... twenty-eighthly. He will stop only after he has overwhelmed you. He is anxious to remain victorious on paper right to the end and to sleep on the battlefield. To sleep is indeed the word, above all for the reader.[89]

This, sadly, is a fairly accurate assessment of Saint-Pierre's otherwise admirable work on the dangers of political dictatorships.[90]

Jean-Jacques Rousseau was more impressed, however, and he revived the Abbé's European scheme in an elegant abridgement in 1761.[91] As far as Rousseau was concerned, European unity was already established at one level:

there are not today any more Frenchmen, Germans, Spaniards, even English. There are only Europeans. They all have the same tastes, the same feelings, the same customs because none has received a national shape by any exclusive institutions. . . . They are at home wherever there is money to steal and women to seduce.[92]

A more positive form of European unity was, Rousseau believed, unlikely to develop without a major revolution.[93]

The 'Russian question' was ultimately unresolved and the debate concerning Russia's European credentials during the eighteenth century has continued to the present. In a persuasive recent account, Larry Wolff has argued that Enlightenment (particularly French) debates about civilisation and barbarism firmly excluded Russia (together with Poland, Hungary and the Ottoman Empire) from the realm of European culture. It was during the eighteenth century, claims Wolff, that the familiar east–west division of Europe definitively replaced the earlier, north–south division. Whereas the north–south classification was usually articulated in terms of old (south) and new (north) Europe, the former spawning the latter, the east–west division was, Wolff implies, a much more prejudicial and judgemental distinction between modernity (west) and tradition (east). This distinction became a geographical shorthand for the supposed cultural dichotomy between 'backward' and 'civilised' regions. Despite the predominantly Christian faith of its inhabitants, eastern Europe was 'invented' during the eighteenth century as western Europe's new constituting 'Other'; as a barbarous realm beyond the limits of the map of civilisation.[94] Other authorities offer a very different chronology and interpretation of 'east–west' and 'north–south' cultural divisions within Europe. Robin Okey suggests a more complex process whereby an earlier east–west conceptualisation gave way to a north–south imagined division in the sixteenth century. This was

> to dominate public perceptions until the early nineteenth century. . . .
> The period of the north–south axis in European consciousness was . . .
> one in which educated élites increasingly distanced themselves from
> ethnocentric or religious zeal in pursuit of rationalistic models in a
> European state system, a balance of power, a mechanism of progress.

This was in turn replaced by a modern version of the east–west divide in the mid-nineteenth century: 'when the masses began to claim a voice in public life, ideological issues were sharpened, and nationalism focused attention once more on cultural difference'.[95] Despite these differing views, most commentators agree that as the Muslim Ottoman threat declined to political irrelevance, a new external 'threat' was invented in the Christian east against which 'enlightened' western Europe could define itself.

People, nation, empire: the geopolitics of the masses

During the nineteenth and early twentieth century the idea of Europe became even more complex and disputed. The French Revolution was a defining moment in this respect. The modern political landscape owes much of its shape and character to the events which followed 1789. The revolutionary political culture which developed in France after 1789 was based in large part on new ways of understanding time and space. Jules Michelet, the most sympathetic nineteenth-century historian of the Revolution was at his most rhapsodic on this theme: before 1789, he claimed, 'the world was a prison'; after it 'geography itself was annihilated. There were no longer any mountains, rivers, or barriers between men. . . . Time and space, those natural conditions to which life is subject, were no more'.[96] Hyperbole perhaps, but the Revolution did give rise to a new decimal calendar dating from the establishment of the Republic rather than the birth of Christ and structured around 'republican' months derived from the natural seasons in place of the old Gregorian sequence.[97] Time and history were of crucial significance to the revolutionaries who saw themselves as breaking with the past, replacing one view of time with another. One wider implication of this was the onset of a recognisably modern form of historical debate, which was to continue through the nineteenth century in the work of writers and pioneering historians such as Michelet, François Guizot, François René de Châteaubriand, Jacob Burckhardt and others, on the meaning of the Revolution within a longer European history. While Michelet saw the Revolution as a crowning glory of European progress, conservative romantics such as Châteaubriand and Burkhardt interpreted 1789 as the final nemesis of the old spirit of Europe which had been in decline since the golden age in the Medieval period.[98]

Space too was imagined afresh after 1789. As Mona Ozouf has observed:

> There is no end to the list of spatial metaphors associated with the Republic or with Revolutionary France: from the beginning of the Revolution a native connivance linked rediscovered liberty with reconquered space. The beating down of gates, the crossing of castle moats, walking at one's ease in places where one was once forbidden to enter: the appropriation of a certain space, which had to be opened and broken into, was the first delight of the Revolution.[99]

The re-organisation of the French administrative system was but one manifestation of the new spatial order. Beginning in the 1790s, the territorial units of the *Ancien Régime* were swept aside. Into their place came a new, rational political space based on a larger number of small *départements*, each administered by a state official, the *préfet*, nominated by the central government in Paris [Figures 1.11(a)–(b)]. The borders of the new *départements*, though modified to take into account natural features, were

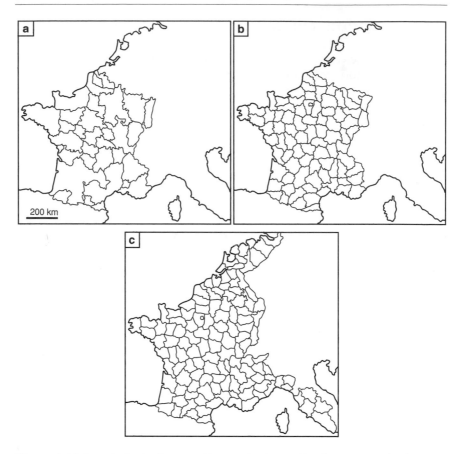

Figure 1.11 Revolutionised space: France in (a) 1789, (b) 1791 and (c) 1812. Source: partly afer P. Jones (1988, 22–3).

initially based on a simple Cartesian grid laid over a map of the national territory.[100] The objective was to enhance the power of the revolutionary core at the expense of an ideologically suspect periphery where regional *Parlements* dominated by local aristocrats and Catholic clerics had posed an obvious threat to central authority (a threat which was to manifest itself in bloody counter-revolutionary wars across western and southern France during the mid-1790s). For the revolutionaries, Paris was to control the national space as the heart controls the workings of the body. Paris alone was to supply the revolutionary energy which would sustain the entire nation, and the geography of the revolutionary state had to reflect that simple fact.[101]

The centralisation of French political power was scarcely unprecedented, of course, but the new concepts of political and administrative space which developed after 1789 intensified this process. Administratively and geopolitically, the First Empire extended the revolutionary spatial order to other parts of Europe. By 1812, the zenith of Napoleonic power, the entire map of Europe had been transformed and new 'French' *départements* stretched across the north European plain east of the Elbe and southwards as far as Rome [Figure 1.11(c)].[102] Beyond this stretched a new European political geography of confederations and buffer states dominated by an all-conquering Napoleonic France [Figure 1.12(a)–(c)].

Paraphrasing Alexis de Tocqueville, D.M.G. Sutherland has argued that 'The First Empire was no mere episode in French history . . . beneath the spectacular details of Napoleon's biography the period witnessed the completion of the millennial process of centralization'.[103] These words were written with respect to France alone but they might equally refer to Europe as a whole, for Napoleon's domination of the continent was based on a bureaucratic and governmental revolution designed to make this sprawling empire function as an integrated political whole.

As Balzac once observed, 'To organise is a word of the Empire'.[104] The Napoleonic Empire raised state organisation and administration to new levels of intensity and sophistication. It is ironic that while the First Empire is firmly associated in the modern historical imagination with romantic military campaigns, sweeping cavalry charges and dashing young officers, its fleeting success probably owed more to the horn-rimmed bureaucrat at his high chair, armed with nothing more dangerous than ledger, quill pen and ink pot. The Empire's most enduring legacy may have been the transformation of political power and statecraft into a 'social science' based on an information revolution of statistics, censuses and endless government inquiries.[105]

The revolutionary and Napoleonic period strengthened the idea of Europe, not least by spawning a richer and more complex historiography of the continent. It also witnessed the first concerted attempt to unite Europe culturally and politically.[106] But it has left an ironic and paradoxical legacy. A united Europe, so often invoked before and since 1789 as a means of assuring peace and harmony between different peoples, came momentarily into being only as a result of warfare and revolution. The Revolution spawned the first 'people's armies' made up of conscript soldiers and *sans-cullottes* who fought within and beyond France's 1789 borders to protect the Republic from internal 'subversion' and external monarchist threat.[107] Once established, this military machine quickly developed its own dynamic and, under Napoleon, seized control of the entire political process – the apotheosis of 1789 or its complete betrayal, depending on your political persuasion. Thus was born the first modern military dictatorship in which all state institutions were directed towards a single objective – the military glory of France. Military statistics, so carefully maintained, speak eloquently. In 1789 the French army numbered fewer than 165 000 men; by the summer of 1793, after the *levée en masse* in which all men aged between 18 and 25 were required to serve for a limited period, there were around

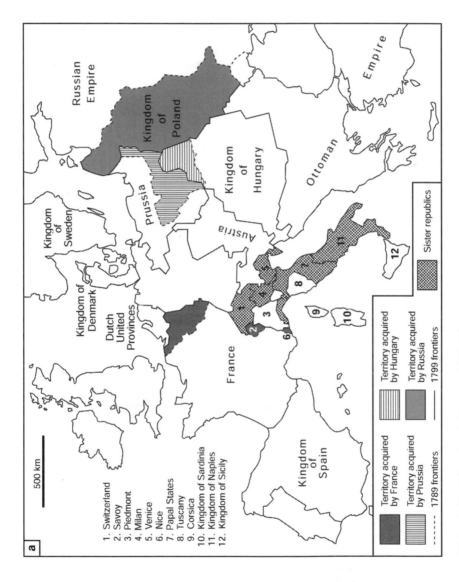

Figure 1.12a Napoleonic Europe: 1789–99

500 km

1. Switzerland
2. Savoy
3. Piedmont
4. Milan
5. Venice
6. Nice
7. Papal States
8. Tuscany
9. Corsica
10. Kingdom of Sardinia
11. Kingdom of Naples
12. Kingdom of Sicily

Territory acquired by France

Territory acquired by Prussia

1789 frontiers

Territory acquired by Hungary

Territory acquired by Russia

1799 frontiers

Sister republics

Russian Empire

Kingdom of Poland

Prussia

Kingdom of Hungary

Kingdom of Sweden

Austria

Ottoman Empire

Kingdom of Denmark

Dutch United Provinces

France

Kingdom of Spain

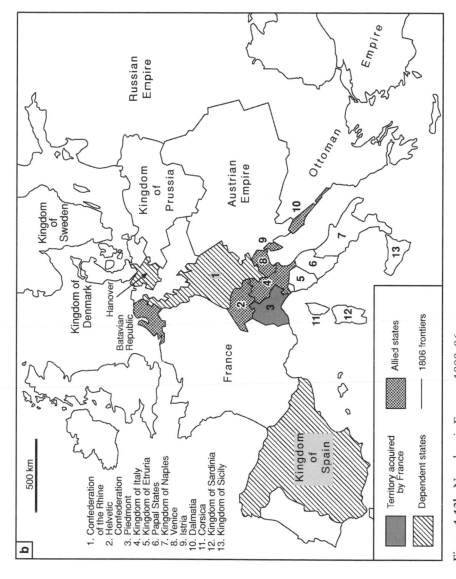

b

500 km

Kingdom of Sweden

Kingdom of Denmark

Hanover

Batavian Republic

France

Kingdom of Prussia

Russian Empire

Austrian Empire

Ottoman Empire

Kingdom of Spain

1. Confederation of the Rhine
2. Helvetic Confederation
3. Piedmont
4. Kingdom of Italy
5. Kingdom of Etruria
6. Papal States
7. Kingdom of Naples
8. Venice
9. Istria
10. Dalmatia
11. Corsica
12. Kingdom of Sardinia
13. Kingdom of Sicily

Territory acquired by France

Dependent states

Allied states

1806 frontiers

Figure 1.12b Napoleonic Europe: 1800–06

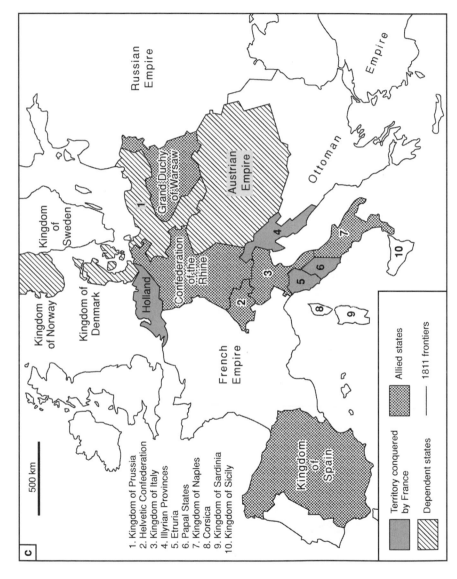

Figure 1.12c Napoleonic Europe 1807–12 Source: after Woolf (1991, 34–6).

1. Kingdom of Prussia
2. Helvetic Confederation
3. Kingdom of Italy
4. Illyrian Provinces
5. Etruria
6. Papal States
7. Kingdom of Naples
8. Corsica
9. Kingdom of Sardinia
10. Kingdom of Sicily

Territory conquered by France

Dependent states

Allied states

1811 frontiers

500 km

Russian Empire

Kingdom of Sweden

Kingdom of Norway

Kingdom of Denmark

Holland

Confederation of the Rhine

Grand Duchy of Warsaw

Austrian Empire

Ottoman Empire

French Empire

Kingdom of Spain

750 000 Frenchmen *sous le drapeau*. By May 1813, in the midst of Napoleon's disastrous Russian campaign which was to destroy his empire, the French imperial armies across Europe probably numbered 1.1 million men.[108] The human cost of this period was enormous. Each year witnessed battles on a larger scale. A 'mere' 40 000 French troops set sail from Toulon to conquer Egypt in 1798; by 1805, 74 000 Frenchmen faced 86 000 Austrians and Russians at Austerlitz, the northern gateway to Vienna. For the ill-fated 1812 invasion of Russia, the Emperor amassed an army of over 600 000 men, only a tiny fraction of whom survived. Overall, two million men served in the Napoleonic armies; 400 000 had died by 1814, and a further 600 000 were registered as missing or prisoners of war. Some 84 000 died in Spain and Portugal; 171 000 in Russia; and 181 000 in Germany. Some 20 per cent of all Frenchmen born between 1790 and 1795 had been killed in battle by 1814.[109] The idea of Europe, in this period as so often later, clearly owed more to war than to peace.[110]

The Napoleonic attempt to create a united Europe was based on a revolutionary and imperial mindset which elevated a national political culture to a European, indeed a universal, level. The idea of Europe as a balance of power between rival states had no place in the Napoleonic order; this was a Europe forged in the image of a single, revolutionary and imperialist state whose civilisation would henceforth speak for all, whose culture was deemed to possess inherent and unquestionable benefits for all men and women. French revolutionary civilisation could, indeed should, be imposed throughout Europe, and ultimately the world, for the good of all humankind. Thus was the idea of Europe conflated with the idea of France and thus were acts of imperial conquest and domination turned into acts of liberation.

In the midst of these epoch-making events, new schemes for universal peace appeared, including Jeremy Bentham's *Plan for an Universal and Perpetual Peace* (1789), the fourth and final essay in his *Principles of International Law* (1780–9) (the work in which Bentham first coined the word 'international'), and Immanuel Kant's *Zum Ewigen Frieden* (1795).[111] These were, however, universal rather than specifically European projects and it was not until the first collapse of the Napoleonic regime in 1814 that a fresh vision of European unity was proposed, the first and most important of the nineteenth century. Its architect was the utopian philosopher Claude Henri de Saint-Simon. His project for a 'reorganised' and united Europe, based initially on an Anglo–French union, was radically different from preceding proposals. Geopolitically, its most important feature was its emphasis on the power of modern technology, transport and communications. Foreshadowing Marx, Saint-Simon felt that the old European nation-states were destined to wither away. Into their place would come a new European political space which would be determined not by the balance of power between nation-states (he was scathing about the Treaty of Westphalia which had made war 'the normal condition of Europe') but by the power of science and technology to create an integrated economic union.

Governments would ultimately be no more than administrative agencies, he argued; the real power would lie with an elite class of European *industriels* and scientists.[112] Saint-Simon's managerial view of technology as a means of creating new political spaces, whether national or international, had enormous impact on subsequent writings on European unity[113] and clearly influenced economists such as Friedrich List who emphasised the role of railways in the creation of new political spaces.[114]

The European geopolitical order constructed at the Congress of Vienna (1815) was entirely different from that envisaged by Saint-Simon (Figure 1.13). The new political map was designed to block future French expansionism: the Habsburg Empire was to protect its new northern Italian territories, and Bavaria and Prussia were allocated lands directly bordering France's eastern frontiers to protect central Europe. The new confederation of German states, dominated by the Habsburg Empire and Prussia, was itself designed to add cohesion, and hence security, to the whole of central Europe. The traditional idea of Europe as an arena of nation-states held in a balance of power reasserted itself with a vengeance through the 'Concert of Europe', an international order to be maintained by periodic conferences of the 'Great Powers'.[115] But the forces unchained in 1789 could not be suppressed. Nationalism had become a genuinely mass ideology and had been immeasurably strengthened in all parts of the continent by the Napoleonic experience. Henceforth, popular nationalism was the dominant force shaping the political geography of Europe.

To begin with, there was no obvious contradiction between nationalism and 'Europeanism', not least because the idea of the balance of power seemed to reconcile these two concepts.[116] Indeed, it was the universalist pretensions of French republicanism and imperialism, against which all European nationalists had united, which seemed to pose the greatest threat to the idea of Europe. As Pim der Boer has noted 'in the revolutionary mentality, there was hardly any place for Europe between citizenship of the world and one's own nation'.[117] Edmund Burke, the father of modern British conservatism and a passionate opponent of the Revolution, was thus able to mobilise a distinctly European argument in his attack on French republicanism. Five years after his initial broadside against the Revolution,[118] Burke set forth his vision of Europe which France threatened to destroy:

> At bottom [religion, laws and manners throughout Europe] are all the same. The writers on public law have often called this aggregate of nations a commonwealth. They had reason. It is virtually one great state having the same basis in general law. . . . [N]o citizens of Europe could be altogether in exile in any part of it. . . . When a man travelled or resided for health, pleasure, business or necessity from his own country, he never felt himself quite abroad.[119]

The 'Europeanness' of early nineteenth-century nationalism was evident during the great revolutions of 1848, the 'springtime of nations'. This series

——— Boundary of German Confederation

1. Denmark
2. Hanover
3. Mecklenburg
4. Netherlands
5. Luxembourg
6. Prussia
7. Saxon States
8. Saxony
9. Austrian (Habsburg) Empire
10. Bavaria
11. Württemberg
12. Baden
13. Switzerland
14. Lombardy–Venetia
15. Piedmont–Savoy
16. Parma
17. Modena
18. Lucca
19. Tuscany
20. Papal States
21. Montenegro
22. Kingdom of the Two Sicilies
23. Kingdom of Sardinia
24. Poland (Russian Empire)
25. Bosnia (Ottoman Empire)

Figure 1.13 Europe after the Congress of Vienna (1815). Source: after C.H. Williams (1989, 205).

of political upheavals, spreading from city to city, swept away the last vestiges of the European *Ancien Régime* in the name of liberalism, democracy
and popular nationalism. The fact that these extraordinary upheavals took
place in so many regions at virtually the same time demonstrates that
nationalism, in its youthful phase, was part of a broader European phenomenon. Inspired by the utopian beliefs of Saint-Simon and earlier federalists, many of the ardent young nationalists who temporarily seized power
in 1848 saw themselves as paving the way towards a united Europe of liberal and republican nation-states. By acting together in a common European
programme these new nations would avoid the clash between republicanism
and monarchism which was deemed to have generated the disastrous rise of
French imperialism in the 1790s. Italy was an important focus of this youthful idealism. Giuseppe Mazzini and the members of *La Giovine Italia*
('Young Italy') had been loudly proclaiming since the early 1830s the need
for a unified European order in which a new Italian republic could claim its
rightful place. In their Swiss exile, Mazzini and a handful of other militants
from Italy, Germany and Poland had signed the 1834 'Young Europe' declaration whose slogans 'Liberty, Equality, Humanity', 'Fraternity of
Peoples' and 'Continual Progress' indicate the movement's romanticism and
utopianism as well as the ongoing influence of the more radical aspects of
French revolutionary nationalism. Their slogan, the United States of
Europe, was echoed by nationalists of all political persuasions.

Philippe Buchez, a leading French social Catholic, an admirer of Saint-
Simon and briefly head of the Constituent Assembly in France under the
short-lived Second Republic after February 1848, was one of the principal
French champions of this idea. Buchez's Europeanism was enthusiastically
endorsed by the great French novelist, Victor Hugo. In 1850, Hugo spoke to
all the nations of the continent, insisting that the day would come when,

> without losing your distinctive qualities and your glorious individual-
> ity . . . [you] will forge yourselves into a close and higher unity: you
> will form a European fraternity, just as Brittany, Burgundy, Lorraine
> and Alsace are united with France. . . . A day will come when these
> two great groupings that face each other, the United States of America
> and the United States of Europe, will join hands across the seas,
> exchanging their goods, their trade, their industry, their arts, and their
> genius, reclaiming the world, colonising the deserts, improving cre-
> ation under the gaze of the creator.[120]

Some 17 years later, Hugo's federalist ardour remained as intense as ever. In
an extraordinary essay, *L'Avenir*, he predicted that the twentieth century
would see 'one European nationality, one unified government, one immense
fraternal association, democracy at peace with itself'. Paris, needless to say,
would be the capital of this European utopia.[121] Pierre Joseph Proudhon, a
key influence on pre-Marxist European socialism, was also a convinced federalist, famously predicting in 1853 that without a united Europe the great

powers would continue to exploit their workers.[122] The idea of an economically integrated Europe was also promoted by Richard Cobden, the English champion of free trade. Cobden looked forward to a European system of nations trading freely with each other, unfettered by tariffs and other constraints. His campaign, motivated by a genuine opposition to war, was launched amidst much fanfare by a triumphant progress around the courts and ministries of Europe in 1846–7.[123]

Despite Cobden's efforts, the urge to settle national differences by resort to warfare proved irresistible. By the mid-1850s three of the main European powers (Britain, France and Russia) were embroiled in the Crimean War. This conflict, though fought over rival imperial interests in the Ottoman Empire, reinforced the 'east–west' division of the continent by pitting the two 'western' powers against Russia. As Sheehan puts it, '[b]y destroying the post-1815 great power consensus the Crimean War opened the floodgates to aggression and territorial revision in Europe'.[124] Anglo–Russian rivalry in Asia continued through the latter half of the nineteenth century, reinforcing the idea of Russia as an 'alien', Asiatic power.

Over the next 20 years each of the five major European powers – Britain, Russia, France, Prussia and Austria – were drawn into at least one war; France, Prussia and Austria enduring three serious conflicts. These jolts to an already unstable system unleashed an evermore aggressive and acquisitive form of European nationalism. Piedmont rose against Austria (with the support of France, whose troops occupied Savoy and Nice as a 'reward'), initiating the *Risorgimento*, the process of Italian reunification in which the charismatic nationalist leader Giuseppe Garibaldi, a former member of Mazzini's *La Giovine Italia*, acquired his legendary status through his dramatic victories in Sicily and Naples (Figure 1.14). In 1866, Venice was ceded by a beleaguered Habsburg Monarchy to the new Italy, and Prussia defeated Austria to assume unrivalled dominance over the confederation of German states in central Europe. Henceforth, the Habsburg dynasty of Austria and Hungary was bent on compensatory imperial expansion in the Balkans and the northern fringes of the crumbling Ottoman Empire, a region which had been increasingly vulnerable to rival Great Power intervention since the Congress of Vienna in 1815.

Europe's optimists watched these events with understandable dismay. They retained, however, a remarkable capacity to imagine a golden future arising, Phoenix-like, from the ashes of each crisis. This was true even of the Franco–Prussian war of 1870 which dramatically transformed the European balance of power in favour of Germany. The newly established German Empire sealed its military victory with the long siege of Paris through the winter of 1870–1 followed by acquisition of the two easternmost French provinces, Alsace and Lorraine. The new united Germany represented the political culmination of the economic links which had bound together the Germanic principalities and kingdoms since the establishment of the *Zollverein* in 1834. The progression from a German customs union,

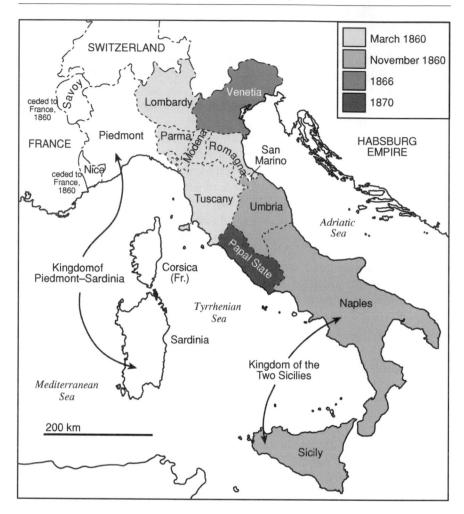

Figure 1.14 The birth of Italy: the *Risorgimento*, 1859–70. Source: after Grenville (1986, 228).

a scheme associated with Friedrich List, to a united Germany was by no means a peaceful process but it was widely seen as an object lesson for a future European union. The intellectual inheritors of Saint-Simon proposed schemes for regional economic co-operation and a wider European customs union on the German model, citing the absurdity of creating evermore efficient means of transport and communications only to stifle these by the imposition of protectionist national tariffs. The 1879 project by the French economist Guido de Molinari for a 'central' European customs union to include France, Belgium, Holland, Germany, Denmark, Austria and Switzerland was but one example.[125] Legal experts also attempted to devise new systems of international sovereignty and justice which might overcome the divisiveness of late nineteenth-century nationalism. The Swiss jurist J.K. Bluntschli offered a scheme for European unity, based on a German core, in his *Europa als Staatbund* of 1871.[126] The great Scottish lawyer, James Lorimer of Edinburgh University, proposed a more radical version of a united Europe in 1884, with a two-chamber Parliament in Geneva.[127]

This enduring optimism, based in part on the comforting idea that a new Europe might emerge from rival states just as a new Germany had arisen from it constituent principalities and kingdoms in 1871, was obviously and tragically misplaced. As we shall see in the next chapter, although nationalism was still consistent with *inter*nationalism during the early and middle years of the century, it evolved into a divisive and confrontational force after 1870. The idea of a harmonious Europe comprising free (and generally republican) nations linked by a common commitment to economic liberalism was already a pipe dream by 1880. Cobden's messianic faith in industrialism and free trade between nations as the harbingers of peace and progress lay in tatters. European nationalism contained within itself the seeds of its own transformation from a pan-European ideology to a competitive and disputatious force which ultimately undermined all sense of a common liberal European consciousness. Once again, the brutal realism of Bismarck was prophetic. When Sir Andrew Buchanan, the British Ambassador to Berlin, spoke to him in 1863 of 'Europe', Bismarck asked a simple but revealing question: 'Who is Europe?'[128]

Summary

This chapter has offered an account of four different ways in which Europe was conceptualised from the Renaissance to the late nineteenth century, each associated with a different form of political territoriality. Although these are presented in a broadly chronological sequence it is important to emphasise that each of these perspectives shaped the debate about Europe throughout this entire period. Older ideas were rarely completely displaced by new ones. The result was a kind of intellectual palimpsest of competing geopolitical visions; a layering of differing meanings.

Notes to Chapter 1

1 P. Burke (1980); Schmidt (1966); Talmor (1980).
2 Dilke *et al.* (1987); Dion (1977); Glacken (1967); Romm (1992); Staszak (1995).
3 J. Fischer (1957).
4 der Boer (1993, 19).
5 Harvey *et al.* (1987); Woodward (1985).
6 Hay (1980).
7 der Boer (1993, 29–32); Brincken (1973); Derolez (1968).

 8 Bartlett (1993); Gollwitzer (1951); Hay (1968); J.W. Meyer (1989).
 9 Southgate (1993, 131).
10 Hay (1980, 1); Rubin (1992).
11 Balzaretti (1992); Leyser (1992); Reuter (1992).
12 Heater (1992, 1–14).
13 Edgerton (1987).
14 Bartlett (1993, 306).
15 Bartlett and Mackay (1989); J.E. Taylor (1993).
16 Manners (1997).
17 Tribe (1978).
18 M. Mann (1986).
19 Brauer (1995).
20 Harley and Woodward (1992; 1994).
21 Quoted in Roberts (1985, 200).
22 E.L. Jones (1987); Dodgshon (1987).
23 Baechler *et al.* (1988).
24 Blaut (1993).
25 Ó Tuathail (1996, 3).
26 Conley (1996); Cormack (1991); Lestringant (1994).
27 The reasons why *Europa* came to be used to denote this region of the earth's sur-
    face is not entirely clear. *Europa* was the daughter of the Phoenician king in
    Greek mythology who was kidnapped and swept off to Crete by a lustful Zeus
    disguised as a bull.
28 Elliot (1992); Fox (1992); Hechter and Brustein (1980); M. Mann (1984);
    Pounds and Ball (1964); Shennan (1974); Tilly (1975; 1989); Wallerstein (1974;
    1980).
29 Piveteau (1995, 15–31); P.J. Taylor (1996, 20).
30 Mukerji (1997).
31 Wintle (1996; 1999).
32 Akerman (1995); Buisseret (1992); Harley (1988); Kain and Baigent (1992).
33 Hertslet (1875–91); Konvitz (1987, 33).
34 Pelletier (1990); see also Dion (1947); Guenée (1986); Nordman (1986);
    Nordman and Revel (1989); Pounds (1951; 1954); Sahlins (1990); N. Smith
    (1969); Weber (1986).
35 Beloff (1967); Sheehan (1996); Wright (1975).
36 Maurseth (1964); Nelson (1943); Pocock (1975); Skinner (1981); Sullivan
    (1973); Vagts (1948).
37 Dionisotti (1971).
38 Parker and Smith (1978).
39 Bull *et al.* (1990); Cutler (1991); P.J. Taylor (1996, 95–8); Vagts and Vagts
    (1979).
40 P.J. Taylor (1996, 21); see also Gross (1948).
41 Sheehan (1988).
42 Hay (1968, 116).
43 Schmidt (1966, 178).
44 P. Burke (1980); Hazard (1990).
45 Hay (1968, 118–19); Outram (1995, 65).
46 Barbiche (1978); Buisseret (1968).
47 Buisseret (1984); Heater (1992, 15–38); Najam (1956); Ogg (1921);
    Soueleyman (1941).
48 Ogg (1921, 25).
49 Ogg (1921, 41).
50 Ogg (1921, 40–1).
51 Neumann and Welsh (1991).

52 Ogg (1921, 31–2).
53 Marshall and Williams (1982).
54 Livey (1981).
55 Agnew and Corbridge (1995, 52–6).
56 Glacken (1967, 551–654).
57 Grosrichard (1979); Springborg (1992).
58 Bauman (1985); Beneton (1975); Elias (1994); Febvre (1974); Kroeber and Kluckhohn (1952); Vogt (1996); R. Williams (1976).
59 Dockès (1969).
60 Nisbet (1980).
61 Colley (1992).
62 Foucault (1991).
63 M. Mann (1993).
64 P. Burke (1992).
65 Hume (1994).
66 M.S. Anderson (1970; 1976); Gilbert (1951); Maurseth (1964); Nathan (1980); Sheehan (1996, 97–120).
67 Sheehan (1996, 97–120).
68 Black (1994).
69 A.W. Fisher (1970); Hosking (1997, 75–149); Jewsbury (1976); Moon (1997); Pallot and Shaw (1990); Sumner (1973).
70 Bassin (1991).
71 Bassin (1988); Slezkine (1994).
72 France (1985).
73 Davies (1996, 8).
74 Bassin (1991, 9); see also Cahnman (1952); Halecki (1952); Hartley (1992); Hosking (1997, 3–41); Kristof (1968); McNeil (1964); A. Palmer (1970); W.H. Parker (1960; 1968); Pesonen (1991); Shaw (1996); Tazbir (1977); Thaden (1984); Wandycz (1992); Wanklyn (1941); Webb (1952).
75 Wildes (1977).
76 Penn (1936, 5).
77 Penn (1936, 6); Heater (1992, 54).
78 Penn (1936, 16); Heater (1992, 55).
79 Penn (1936, 19).
80 Penn (1936, 22).
81 Penn (1936, 30).
82 Bellers (1710); Clark (1987, 132–53); Fry (1935, 89–103); Heater (1992, 57–60).
83 Clark (1987, 140); Heater (1992, 58).
84 Clark (1987, 152); Heater (1992, 59).
85 de Saint-Pierre (1712); Perkins (1959).
86 Heater (1992, 66).
87 Cutler (1991).
88 Heater (1992, 84–5).
89 Quoted in Drouet (1912, 340) and Heater (1992, 66).
90 de Saint-Pierre (1728).
91 Rousseau (1761); Nuttal (1927); Roosevelt (1990, 185–220).
92 Joll (1980, 11–12); Heater (1992, 79).
93 Wokler (1995, 70).
94 Wolff (1994); M.S. Anderson (1954); Chiriot (1989).
95 Okey (1992, 110–11); Stirk (1996, 1).
96 Quoted in H. White (1973, 151).
97 Baczko (1984).
98 Bann (1984); H. White (1973).

 99 Ozouf (1988, 126).
100 Konvitz (1987, 44–5).
101 Konvitz (1987, 32–62; 1990); Margadant (1992); Nordman and Vic-Ozouf
    Marignier (1989a; 1989b); Vic-Ozouf Marignier (1989); Vovelle (1993).
102 Broers (1996, 174–5).
103 Sutherland (1985, 438).
104 Woolf (1991, 83).
105 Le Bras (1986); Perrot and Woolf (1984); Woolf (1989a).
106 Broers (1996); Woolf (1989b; 1991; 1992).
107 Cobb (1987).
108 Woolf (1991, 156–65).
109 G. Parker (1995, 208).
110 J.S. Levy (1983); Midlarsky (1975).
111 Heater (1992, 93–4); Hinsley (1967); von Raumer (1953).
112 de Saint-Simon and Thierry (1814); Ionescu (1976, 83–98); K. Taylor (1975,
    129–36).
113 d'Eichthal (1840); Pecqueur (1842).
114 List (1904); Henderson (1983).
115 Albrecht-Carrié (1968); Hinsley (1966); Langhorne (1981); Midlarsky (1981);
    Moul (1985); Schroeder (1989); Sheehan (1996, 121–44); Stall (1984).
116 Renouvin (1949).
117 der Boer (1993, 68).
118 E. Burke (1983).
119 Quoted in der Boer (1983, 66–7).
120 Quoted in Duroselle (1990, 324).
121 der Boer (1993, 77).
122 Duroselle (1990, 304); see also Proudhon (1979); Newman (1983).
123 Pick (1993, 19–27); P.J. Taylor (1996, 98–102).
124 Sheehan (1996, 132).
125 Stirk (1996, 9–10).
126 Bluntschli (1871).
127 Lorimer (1884); Harvie (1994, 40–2).
128 Duroselle (1990, 333).

|2|

Who is Europe? Fin-de-siècle *geopolitics*

Introduction

This chapter examines debates about Europe's nature, extent and internal geopolitics from the 1880s to the end of World War 1. This troubled period, which ended with the horrors of industrial warfare and the wholesale redrawing of the continent's political map, witnessed the climax of European political territoriality, the culmination of the process outlined in Chapter 1. This was the era in which modern geopolitics was born.

The best of times? Paris in the spring

On 14 April 1900, Europe's rich and famous gathered in an affluent district of west-central Paris to witness the opening ceremony of the *Exposition Universelle*, a festival to the achievements of the nineteenth century and the prospects of the twentieth. The *quartier* had been substantially rebuilt for the event. On the right bank of the Seine loomed an immense triple archway, the entrance to the Exhibition site. Beyond was the Grand Palais with its domed roof of steel and glass flanked by bronze chariots. On the left bank, in the shadow of the Eiffel Tower, itself only a decade old, stretched the Rue des Nations, a phantasmagoria of architectural styles where the 'major nations' had constructed pavilions to house their exhibits. Here was a copy of Washington's Capitol Hill, a German *schloß* (complete with beer garden), an Elizabethan manor house and a replica of part of the Kremlin. Linking the two sites was the glittering span of the Pont Alexandre III, a spectacular neo-Baroque bridge named in honour of the late Tsar and a symbol of the recent alliance between France and Russia.[1]

The *Exposition* was billed as a celebration of international co-operation and harmony. It was, however, a consciously European event. Seven years earlier, the infant city of Chicago had staged its own lavish Exhibition to commemorate the quatercentenary of the 'discovery' of the Americas. The 1900 *Exposition* was Europe's response to this hugely successful American venture, a restatement of the continent's enduring pre-eminence. Throughout that memorable summer, Paris hosted the Olympic Games, second in the modern sequence, together with 130 international congresses, debating everything from socialism to vegetarianism. Science and technology, the liberating forces of the new century, were given pride of place. Motorcars and experimental flying machines were displayed at Vincennes; moving photographic images were projected onto a giant, 360-degree screen; and a Palace of Electricity threw its powerful light, the force of the future, into the Parisian sky. Also popular was an enormous globe, designed by the geographer Elisée Reclus, on which was displayed the earth's continents and oceans. Visitors could glide across the surface from pole to pole in chairs suspended from a spiral encasement.[2] Thus was the new century celebrated in Paris; here, the 'capital of the nineteenth century' asserted its leadership of the twentieth.

A ride around the Reclus globe or a promenade along the Rue des Nations provided ample justification for this confident display of European hegemony. Europe administered more than half the earth's land surface (Figure 2.1). The British Empire alone covered a quarter of the world and encompassed 400 million people. The internal political geography of Europe reflected the 'core ideology' of imperialism. Despite profound constitutional differences between the main European nations (from liberal, republican France to autocratic, Tsarist Russia) all were empires, either within Europe or beyond (Figure 2.2). The Paris *Exposition* of 1900 projected an image of the world fashioned by Europeans united in a common project begun 400 years earlier and destined to continue into the new century.

Not everyone shared this optimism. Letters in the Parisian press suggest a high level of popular scepticism. Some objected to the noise and dirt of the construction work, the ignorance and drunkenness of the tourists and the nefarious activities of the pickpockets and prostitutes they attracted. One correspondent worried that carpets imported from the Middle East for the national pavilions would be infested with plague. Other critiques were more substantial and informed. The *Exposition*, claimed the pessimists, was a tawdry and meretricious display; a despairing attempt to paper over the widening cracks in French, and European, political culture. 'The atmosphere hanging over the banks of the Seine was a strange one', claims one historian, 'an uneven mixture of bombast, pride, fear, insecurity and confusion'.[3]

These contradictory feelings were entirely characteristic of the European *fin-de-siècle*, a term coined in France during the 1880s to capture the ambivalent public response to the century's close. In most countries,

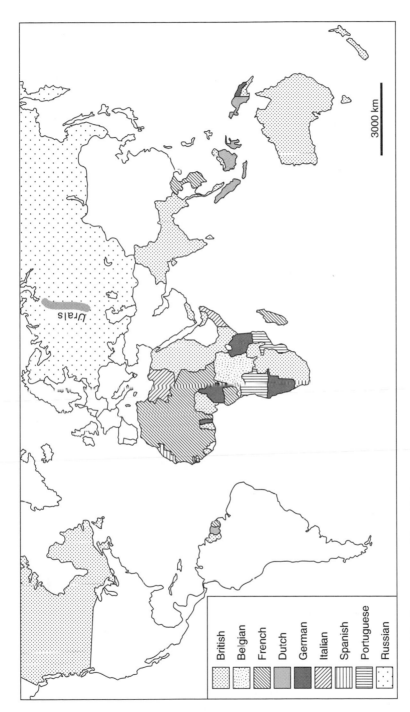

Figure 2.1 The European world, c. 1914.

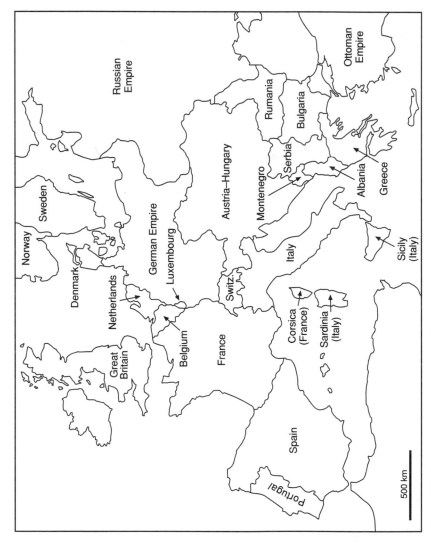

Figure 2.2 Europe on the eve, July 1914.

optimists seemed to be outnumbered by pessimists; 'mourning the twilight' was a more common reaction than 'celebrating the dawn'.⁴ In France, the last three decades of the nineteenth century, beginning with the *annus horribilus* of 1870–1, had provided ample reason for sombre reflection.⁵ The governments of the Third Republic were notoriously unstable and often corrupt. Public confidence in the political process was generally low, and anxiety about the future persistent. The size of the French population, always a crucial index of national well-being, was worryingly static.⁶ The Dreyfus Affair (a bitter crisis prompted by the wrongful imprisonment of a Jewish army officer on trumped up charges of passing military secrets to the Germans) cast a shadow over the 1890s, polarising an already embittered public opinion. Many hoped the new century would bring an end to this painful era, but few looked forward with genuine optimism.⁷

The future seemed no less worrying in London, epicentre of the nineteenth century's dominant European power. The British capital had staged its own spectacle of imperial grandeur in 1897 to celebrate the diamond jubilee of Queen Victoria, Empress of India and the monarch whose name symbolised the era of British hegemony. Victoria died in 1901, a demise which seemed poetically to confirm the closing of 'the British century'. The outbreak of the Boer War in South Africa in 1899 reinforced mounting fears that Britain could no longer bask in the serenity of an untroubled empire. The suppression of Afrikaner resistance cost 70 000 lives, £20 million and undermined Britain's reputation as a benign, unassailable imperial power. Britain, claimed Joseph Chamberlain in 1902, was a 'Weary Titan that staggers under the too vast orb of his own fate'.⁸

A sense of foreboding was equally evident in other European capitals. Vienna and Budapest, twin citadels of Austria–Hungary, looked out over a vast empire wracked by ethnic tensions and political instability.⁹ In St Petersburg the quasi-medieval Tsarist regime clung tenaciously to the most ancient forms of autocratic privilege. The Romanov dynasty was rocked by the unexpected defeat of its imperial armies at the hands of the Japanese in 1904 and by the subsequent struggle for democracy during the revolutionary upheavals of 1905. The establishment of a new Russian *Duma* (Parliament) was an inadequate response to the simmering resentments which were to erupt so spectacularly in 1917.¹⁰ Across the continent, the *fin-de-siècle* seemed to herald a wider crisis: the *fin de l'Europe*.

This was no idle speculation, for the years between *c.* 1890 and the outbreak of World War 1 witnessed a profound transformation in the nature of the world system which was to leave Europe more vulnerable than at any time since Columbus first set sail.¹¹ The context was a shift from an industrial capitalism based on steam, coal and iron to a new age dominated by gas, oil and electricity, coupled with an increasingly international finance capital. The economy of the USA, poised on the brink of a 'Fordist' revolution of intensive industrial production for a mass market, was perfectly placed to dominate the new century. By the eve of World War 1, US factories were producing a third of the world's industrial goods, a dominance which Britain had claimed a mere four decades earlier. Viewed from this economic perspective, the 1893 Columbian Exhibition in Chicago was a more fitting celebration of the coming century. America was the future; Europe the past.

The emergence of the USA as a continental-scale power threatened the very basis of the European geopolitical system with its rival nation-states and distant, maritime empires. This system had facilitated Britain's global dominance but the rise of the USA changed the ground rules. Many believed that only a united continental Europe could mount a serious challenge to the USA in the twentieth century. Only one European power, Germany, had the potential to bring this about. Unlike the older European industrial economies, Germany's development was based on the new forms of industrial product. The German economy had grown at an extraordinary rate since unification in 1871, despite periodic crises of overproduction. By 1913, Germany's productivity far exceeded Britain's and was surpassed only by that of the USA. Needless to say, the prospect of a German Europe was extremely worrying for the other European powers. The rapid rise of Germany, and its bid for global supremacy after 1890, generated a surge of competitive nationalism, a 'geopolitical panic' which transformed and destabilised the European system. There were three related dimensions to this crisis.

First, Europe's commerce was increasingly undermined by an economic nationalism which destroyed the free-trade idealism so hopefully discussed in the 1840s. The European clamour for tariff reform and protectionism, and the associated belief in autarky (self-sufficiency), can be seen as a general response to the first stirrings of a global economic system which national governments could do little to control; it was old-style political nationalism striking back against capitalist internationalism. But it was also connected to internecine European rivalries, particularly the fear of Germany. Bismarck inaugurated a more protectionist economic policy in the late 1870s, and successive French governments responded in kind, culminating with the Méline tariff of 1892 which sealed off the French agricultural sector from the rest of Europe. Other countries followed suit, leading to a series of 'tariff wars' during the last years of the nineteenth century.

Second, the European nations embarked on a new phase of imperial expansion, a 'scramble' for colonial territory described by the French politician Jules Ferry as a 'steeplechase into the unknown'. Over 16 million square kilometres and 150 million people were added to the European empires during the last three decades of the nineteenth century, particularly in Africa, which was divided between the great powers at the Berlin conference of 1884–5. The new imperial territories represented 20 per cent of the earth's surface and 10 per cent of the world's population. On the face of it, imperial expansion seemed to underscore Europe's continuing hegemony. But the 'new imperialism' of the 1880s and 1890s was rather different from earlier episodes of European expansion. Although legitimised by the age-old rhetoric of civilisation, commerce and religion, the yearning for new colonial space after *c.* 1880 sprang from an anxious and defensive European nationalism. Precise objectives varied, of course, but *fin-de-siècle* imperialism was often determined by fear that 'unclaimed' colonial territory would fall into the hands of rival powers. Colonies were coveted not for economic reasons but as symbols of an otherwise vulnerable national pride, tangible zones which could be coloured in the appropriate fashion on school atlases, wall maps, tea towels and a thousand other items of popular imperial propaganda.[12] No self-respecting 'great power' could be without its 'place in the sun'.[13]

Many late nineteenth-century European empires barely broke even financially. Only 10 per cent of French exports and an even smaller percentage of overseas investment was directed towards France's empire. Less than half the exports from French colonies ended up in *la mère patrie*.[14] The German *Kolonialreich* in Africa and Asia cost the 'Fatherland' £6 million in 1913.[15] *Fin-de-siècle* imperialism was not so much 'the highest stage of capitalism', as Lenin described it in 1916, as the climax of a defensive European nationalism. Empire, so easily interpreted as a sign of Europe's power and unity of purpose, could equally be viewed as a symbol of its divided weakness.

The third dimension to Europe's 'geopolitical panic' involved a fundamental reordering in the system of alliances. Prior to the 1890s, the interstate system had been an ever-changing kaleidoscope of bilateral treaties and understandings. The only persistent features were antagonism between France and Germany and the alliance between Germany and Austria–Hungary. Paradoxically, this very complexity, coupled with Britain's 'splendid isolation' from European intrigues, produced a relatively stable system in which wars, though by no means infrequent, could at least be contained. The rise of Germany made the European system less stable.

The shift took place in two phases which transformed the foreign policies of first Russia and then Britain, locking both empires more firmly into the European system. During the 1890s, a growing body of opinion in St Petersburg argued that the new Germany was destined to seek expansion in the east and therefore posed a direct threat to Russia. This led to an alliance between Russia and France, inaugurated by a series of loans and economic accords and secured by a formal military convention in 1894. The objective was to enclose Germany from east and west. This prompted equivalent moves in central Europe to cement the 'Triple Alliance' between Germany, Austria–Hungary and Italy. Three 'power blocks' now structured the European geopolitical order: the British Empire, the Franco–Russian alliance and the central European 'Triple Alliance'.

The second phase, involving Britain, converted this tripartite structure into a more dangerous, bipolar, system. Like their Russian contemporaries, British diplomats and politicians calculated that their imperial interests in the Middle East, the gateway to British India, were directly compromised by

German plans for an economic drive into the vulnerable Ottoman Empire, sometimes called the *Drang nach Süden* (the 'drive to the south'), a geopolitical objective symbolised by the planned railway link between Berlin and Baghdad. Britain's fear of Germany overcame the traditional distrust of France, and an *Entente Cordiale* was signed in 1904 settling the outstanding disputes between London and Paris, mainly in the Middle East. Three years later, Britain and Russia, whose previous relationship had been equally tense, signed a further accord which sought to end their long-standing imperial rivalry in Asia. A 'triple entente' between Britain, France and Russia now surrounded the central European 'Triple Alliance', designed to complete the containment of Germany.

The emergence of a bipolar European system may well have been inevitable given the impossibility of reconciling the territorial demands of rival imperial states. What this produced, however, was a deepening in international tension and a huge expansion in the military capacity of each state. The size of the main European standing armies (Russian, French, German, British, Austro–Hungarian and Italian) grew by an average of 73 per cent between 1880 to 1914. The warship tonnage of these powers increased almost fourfold over the same period, largely as a result of the head-to-head contest between Germany and Britain, the latter expanding its navy from 650 000 to 2 714 000 tons, the former from 88 000 to 1 305 000 tons.[16]

Much intellectual energy was expended inventing ways to deploy these unwieldy military machines. Plans were endlessly updated. The most famous scheme was developed by the German High Command, under the Chief of the Imperial General Staff, Count Alfred von Schlieffen. Confronted by the prospect of war on two fronts, Schlieffen's audacious solution was a massive initial attack against France. Accepting the probability of substantial losses to French counter-attacks in Alsace–Lorraine, and banking on a lumbering Russian attack in the east, the plan foresaw five armies, three-quarters of Germany's military capacity, sweeping in a great arc into northeastern France via Belgium. Paris, the ultimate objective, would be attacked from the north. The aim was a quick victory in this western 'theatre', as in 1870. Troops and weaponry could then be rapidly redeployed towards the east by train to repulse the invading Russian forces. It was hoped that Britain would hesitate before abandoning its traditional policy of 'splendid isolation'. Even if it did not, the German High Command calculated that the small British army would have little immediate impact. Schlieffen's ambitious plan was substantially scaled down by his successor, General Helmuth von Moltke, but its broad objectives were retained as the foundations of German strategy down to 1914.

In 1907 the only international accord which survived from the 'Concert of Europe' and which cut across the bipolar system of alliances was a tenuous Austro–Russian entente. This ended abruptly in 1908 when the Habsburg monarchy annexed Bosnia and Herzegovina, an event foreshadowed by the gradual collapse of Ottoman authority throughout the Balkans under pressure from nationalists. This same impulse precipitated general warfare in 1912–13 which produced a complex mosaic of small, mutually antagonistic Balkan states, the focus of international intrigue as both sides of the bipolar system sought to acquire territory and influence in the region (Figure 2.3). The adjacent Habsburg monarchy of Austria–Hungary was the most threatening and acquisitive power, the 'cat's

Figure 2.3 The cradle of war: the Balkans, 1913–14. Source: after Pounds (1985, 27).

paw' for the Central Alliance, occupying Bosnia and Herzegovina in 1878
and incorporating these lands into the Empire from 1908.

The desire to annex or dominate Balkan territory led directly to the
firestorm of World War 1 (Figure 2.4). The detailed events leading up to
August 1914 are immensely complex.[17] The central point is that the assassination of the heir to the Habsburg Empire, the Archduke Franz Ferdinand,
at the hands of a young Serbian nationalist, Gavrilio Princip, in the Bosnian
capital of Sarajevo in late June 1914 sparked a conflict across the bipolar
divide between Austria–Hungary and Serbia, which was 'protected' by a
Slavic alliance with Russia. The crisis provoked a general European mobilisation as the system of alliances shifted into gear, an escalating process
which has been the subject of considerable debate. Recent interpretations,
inspired by the British historian A.J.P. Taylor, emphasise the powerlessness
of politicians and diplomats; their apparent inability to control the military
machines they had created. The mere presence of new technologies, it has
been argued, particularly in transport and telecommunications, meant that
crucial decisions were made at high speed. As a result, events spiralled out
of control, developing their own, malign dynamism.[18]

In early August, Molke's invasion plan was put into action (Figure 2.5).
Contrary to German hopes, Britain entered the war on 4 August to honour

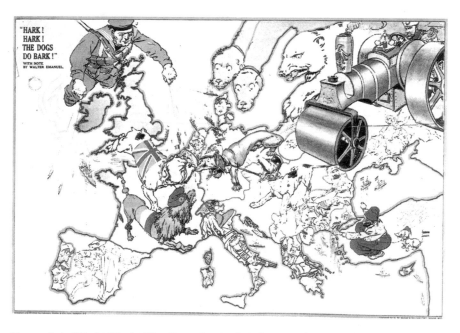

Figure 2.4 'Hark! Hark! The Dogs Do Bark!'. Source: George Washington Bacon,
HMSO. Note: This image, by George Washington Bacon, was originally a postcard
on sale in English seaside resorts during the high summer of 1914.

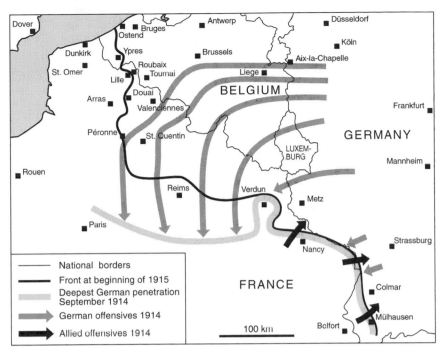

Figure 2.5 August 1914: the German invasion of Belgium and France. Source: after A.J.P. Taylor (1963, 30).

alliances with Belgium and France. After fierce fighting, the German advance through France was halted at the Battle of the Marne, to the north and east of Paris. There followed a German retreat to the River Aisne and the misnamed 'race for the sea' when the opposing armies sought to out-flank each other in the north. Exhausted by this inconclusive collision, both sides began to dig lines of defensive trenches which eventually stretched in a meandering scar from Flanders in the north to the Vosges in the south. Despite relatively minor modifications as a result of the terrible battles around Ypres, Verdun, the Somme and the Chemin des Dames from 1915 to 1917, this line of opposing armies remained a permanent cartographic feature (Figure 2.6). War in the east started disastrously for Russia, whose armies invaded East Prussia in mid-August but were soon repulsed and then heavily defeated at the Battle of Tannenberg. The conflict between German, Austro–Hungarian and Russian forces in the eastern 'theatre' was less static than in the west and, with some notable exceptions, went favourably for the Central Powers.

The war quickly escalated beyond the limits of the European bipolar sys-tem. The Ottoman Empire sided with the Central Powers in November 1914; Italy resisted pressure from its pre-war allies and then took arms against them in April 1915 after secret negotiations in London, where exor-bitant schemes for a post-war Italian empire in the Adriatic and Asia Minor

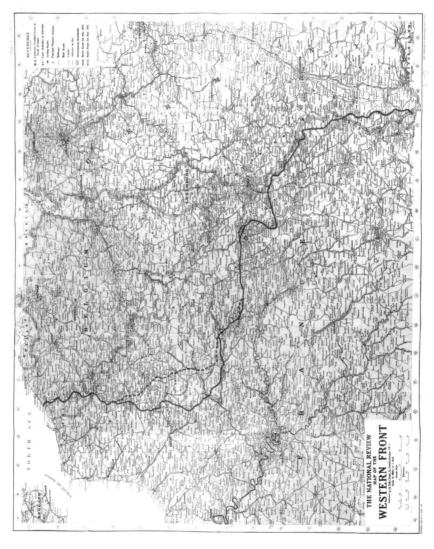

Figure 2.6 The Western Front: *The National Review* map, 1917. Source: Cambridge University Map Library, Maps. c. 23.91.8.

were discussed; Bulgaria allied itself with the Central Powers the following October. These developments did little to break the deadlock and it was not until 1917 that two events transformed not only the course of the war but the entire history and geography of Europe. The first was the entry of the USA into the war on the Allied side in April 1917, a decision precipitated by the German declaration of unrestricted submarine warfare against ships supplying its enemies. The second was the withdrawal of Russia from the conflict following the Bolshevik Revolution towards the end of the year. A separate peace treaty was signed between Germany and Russia at Brest-Litovsk on 3 March 1918. This surrendered huge tracts of former Russian territory, allowing the temporary establishment of an eastern 'empire' of German satellite states, including a new Poland, Ukraine, Azerbaijan, Armenia, Georgia, Finland, Latvia, Lithuania and Estonia. This fleeting political geography, when coupled with Germany's pre-existing war aims as laid out in the September 1914 memorandum issued by the German Chancellor, Theobold von Bethmann-Hollweg, gives us some idea of what a future Europe would have looked like had Germany been victorious (Figure 2.7).

The two main events of 1917 worked in opposite directions: the US entry favoured the Allies, the Russian withdrawal helped the Central Powers. Germany gambled on a final military resolution before US mobilisation could take effect and launched an all-out assault in the west in the spring of 1918. Like the 1914 offensive, this ground to a halt within shelling distance of Paris and was then reversed by the combined Allied forces, with the two million US troops playing an increasingly important role. Facing military defeat coupled with domestic economic collapse and political upheaval after the years of wartime blockade, the German government sued for peace. An armistice was declared on 11 November 1918, the Kaiser abdicated and the Hohenzollern monarchy followed the Romanovs into oblivion.

The years between the Franco–Prussian war of 1870 and the end of World War 1 were profoundly traumatic for the people of Europe. Not surprisingly, these shifts and upheavals were accompanied by widespread, soul-searching debate about Europe's past, present and future. Geographical questions about the continent's place in the global scheme, its physical extent and its internal political geography loomed large. 'Geopolitics', a term coined in 1899 by the Swedish geographer Rudolf Kjellén to describe the 'scientific' study of the state, its borders and its relations with neighbouring states, emerged as a distinctive intellectual project with its own language and rhetoric. 'Geopolitical' theorising about Europe reached a new intensity, both within Europe and around the world. The remaining sections in this chapter consider a selection of the different geopolitical debates about the nature of Europe which were advanced during this troubled era.

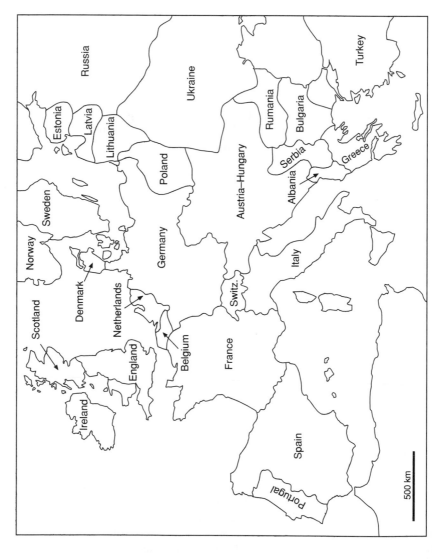

Figure 2.7 Germany's Europe, early 1918. Source: after Blouet (1996, 7); F. Fischer (1967, 102–5); Koch (1984).

The end of Europe: Eurasia and the geographical pivot

The prospect of a world in which all regions were explored, mapped and allocated politically was a central preoccupation of *fin-de-siècle* geography. For optimists, the earth's 'closure' raised the possibility of a new, planetary consciousness, a utopian 'one-worldism'. The 1891 proposal from the renowned German geographer, Albrecht Penck, for an international world map to be based on a 1:1 million scale and a standard set of conventions and symbols illustrates the hopeful and celebratory dimension of 'closed-space' thinking. Four centuries after Columbus, argued Penck, on the eve of the twentieth century and with only the polar regions still unexplored, the time had come for the great powers to unite in a common project to construct a single cartographic image of the world, an international tribute to the age of European expansion and exploration. Enthusiastically endorsed by leading geographers and initially by several governments, this intriguing scheme fell foul of the international tension it sought to overcome. Penck himself was later to repudiate internationalism in favour of a more conservative nationalism.[19]

Other 'closed-space' theorists were more pessimistic about the future. The British geographer and politician Halford Mackinder was an eloquent representative. Much has been written on Mackinder, a towering figure in the early history of British geography, as a university and school subject. He was appointed Reader in Geography at the University of Oxford in 1887, the first modern acknowledgement of the subject's significance by the country's senior university. He went on to become Principal of the fledgling university at Reading, and Director of the London School of Economics (LSE). This was followed by a political career as a Unionist member of parliament after 1910.[20] Although he retained a belief in the value of travel and exploration (he made the first ascent of Mount Kenya in 1899)[21] Mackinder is best known as the leading prophet of the 'new geography' in Britain. Drawing on the ideas of French and German geographers, especially Friedrich von Richthofen, he blazed a trail for a more theoretically informed, synthesising subject, a conceptual bridge between the natural and social sciences.[22] Mackinder was an influential advocate of imperial reform, and his ideas were shaped by his desire to preserve and enhance Britain's position as a global power. His was a planetary political geography, an attempt to understand how the different nations and regions of the world interrelated as elements in an holistic geopolitical structure. Education lay at the heart of his concerns and geography lay at the core of his educational strategy. An imperial nation such as Britain demanded the cultivation of an 'imperial race' trained in the art of 'thinking imperially'. This depended on a spatial and visual sensibility gleaned from map and landscape interpretation, the central preoccupations of the geographer.[23]

Mackinder's desire to build a 'new geography' was informed by his astute recognition that the world itself was changing in fundamental ways. His views were eloquently expressed in a famous lecture, delivered before the Royal Geographical Society in London in January 1904.[24] Like Penck, Mackinder argued that geographical exploration was a finite task which would soon be completed. The age when the basic geographical facts about the world were fully appreciated was at hand. Geography, hitherto driven by the necessary but conceptually straightforward challenge to map and identify the 'unknown' places of the earth, needed to respond to the challenges of the new century: 'geographical exploration is nearly over', Mackinder insisted, 'and . . . geography must be diverted to the purpose of intensive survey and philosophic synthesis'.[25]

Failure to develop explanatory geographical theories carried dire consequences for the 'old' imperial states. The closing of the nineteenth century marked the end of the 'Columbian age' of European expansion, the era when Europeans had escaped their Medieval limits and expanded 'against almost negligible resistances' [*sic*]. Europe's 'colonial frontiers' had now been pushed to their limits: 'there is scarcely a region left for the pegging out of a claim of ownership'.[26] This was a worrying development. Like many imperialists, Mackinder believed that the colonisation of the 'wide-open spaces' of Africa and Asia had acted as a necessary 'safety valve', allowing Europe's restless ambition to be deflected away from territorial conflict in Europe and towards the useful development of empires. The 'closure' of the world to further European expansion would generate social and political tension within Europe while increasing the likelihood of conflict along imperial frontiers which were now shared with rival imperial powers and hence static rather than unfolding. He wrote:

> Every explosion of social forces, instead of being dissipated in a surrounding circuit of unknown space and barbaric chaos, will be sharply re-echoed from the far side of the globe, and weak elements in the political and economic organism of the world will be shattered in the consequence.[27]

Mackinder's Eurocentrism, his refusal even to consider the rights of colonised peoples, was characteristic both of his class and of the era in which he wrote. But his analysis was undeniably correct in one respect. By the late 1890s, European expansion into Africa and Asia had became a source of conflict between the great powers rather than a means of reducing tension. The British desire to establish a north–south 'Cape-to-Cairo' link through east Africa, for example, clashed directly with French objectives to open a route across the continent from west to east. The two countries came perilously close to war in 1898 when Kitchener's troops confronted Marchard's at Fashoda on the banks of the White Nile. Russia and Britain were also embroiled in a long-running 'Great Game' of espionage and counter-espionage, the 'Cold War' of the imperial age, along the flanks of their respective Asian empires in the mountains and foothills of the Hindu Kush and the Kyber Pass. Although these contests would be partially resolved by the emergence of a bipolar European system after 1907, the potential for European crisis across this divide was ever present. France and Germany twice came to the brink of war, in 1905 and 1911, over their rivalry in Morocco.

The end of the 'Columbian age' represented, Mackinder asserted, a crisis for the old imperial powers of Europe, particularly Britain. With a minuscule 'domestic' territory and widely scattered imperial possessions, Britain was vulnerable both to rival imperial powers and to gathering resistance within its empire (as the Boer War demonstrated). 'Imperial overstretch' would be exacerbated, Mackinder claimed, by the eclipse of sea power in the post-Columbian era.[28] Sea power, insisted Mackinder, had been the basis of Europe's hegemony in the 'Columbian age' but global dominance in the future would depend on control and exploitation of the vast resources of the world's land masses.[29] In the new age, a revolution in overland transportation by rail would mean that continents could be bound together by networks of iron, radiating from great transcontinental railways. The Trans-Pacific railroad across North America and the Trans-Siberian across Russia's Asian empire were the first, probing tentacles in a new transport system destined to encircle the globe.[30]

Asia, the largest land mass on the planet, was the key to Mackinder's world-view. This was his 'geographical pivot of history'. The idea that a new railway system could conquer the immense expanse of Asia was an awesome, and deeply disturbing, prospect. 'The century will not be old', he predicted, 'before all Asia is covered with railways'.[31] The failure of this system to materialise does not detract from the provocative impact of Mackinder's prediction in 1904. European civilisation, 'the outcome of the secular struggle against Asiatic invasion' in Mackinder's estimation, was directly threatened by a new Eurasian land empire occupying the geostrategic centre of the globe. Europe's global hegemony, established by small, seafaring nations which had embraced the rule of law and democracy (Britain, France and the Netherlands) had prevailed in the age of sail and steam. The world they created was unlikely to survive into the age of continental railways and long-distance land travel. Asia was poised on the brink of global dominance. The question was, which of the existing states around the edge of the Asian land mass was destined to control this crucial arena?

Imperial Russia, the only existing 'Eurasian' power, was best placed to dominate the Asian land mass. However, this awesome task might prove too much for an unreformed Tsarist Russia which seemed incapable of embracing the modern age (the Trans-Siberian railway notwithstanding). Other powers, located in the 'crescent' surrounding the Asian land mass, offered alternative threats. The emergence of Japan, whose forces were poised to defeat the Russian army as Mackinder was preparing his lecture in 1904, raised the spectre of an Asian empire developing from the east rather than the west. But Germany, a state apparently obsessed with territorial expansion, seemed the obvious non-Russian source of a Eurasian empire. The dread scenario, Mackinder believed, was an alliance between Russia and Germany. This would establish an irresistible force – the modern, machine-efficiency of Germany coupled with the massive demographic bulk of peasant Russia. The Asian land mass, 'the world island', would be swiftly overwhelmed by such an alliance. This would herald an entirely new era in the world's geopolitical system. Whereas previous episodes of European colonial expansion had created peripheral, maritime empires which had sustained the dominance of the European core, the conquest of the 'world island' would represent not the expansion of Europe but its virtual abandonment in favour of the huge expanse of Asia. European civilisation, the maritime empires and the traditional balance of power would collapse. The 'Columbian age' of European dominance would end and a new, Eurasian era would begin. If Britain and Europe were to survive into the twentieth century, the fearful prospect of a Germano–Russian alliance must be prevented at all costs (Figure 2.8).

The historical context within which Mackinder wrote was all-important. His lecture was delivered during the preliminary discussions for the Anglo–French *Entente Cordiale*, signed 10 weeks later. As we have seen, this accord laid the foundations for the 'triple entente' and, through this, the emergence of a bipolar European order, an arrangement secured in 1907 by the Anglo–Russian entente. Mackinder's analysis of the Eurasian threat to the existing European order was an elaboration of Britain's new role in European power politics, an exercise in 'shock' tactics designed to underscore the need for an 'encircling' European alliance which would contain Germany on all sides, bind Russia to France and Britain and remove the prospect of an alliance between the two potential Eurasian powers. The British lion, Mackinder implied, must awaken from its complacent imperial slumber. Any residual belief in Britain's 'splendid isolation' from European political intrigues was no longer tenable. The world had changed, claimed Mackinder, and Britain had to respond to protect its position and to ensure the survival of European civilisation. Mackinder's 1904 lecture was a defining moment in the history of geopolitics; a foundation statement about the dangers to Britain and Europe in the twentieth century. It was both a reflection of, and a contribution to, a global territoriality which has grown steadily stronger during the twentieth century. For Mackinder, and those who followed his reasoning, geographical size was all-important. Land, resources and 'manpower' (a term Mackinder later claimed to have coined) were assumed to determine national significance and prestige.[32]

In his *Democratic Ideals and Realities: A Study in the Politics of Reconstruction*, published on the eve of the post-war Peace Conference in early 1919, Mackinder offered an extended analysis of his 1904 paper and a pessimistic interpretation of Europe's future.[33] The war had 'established, and not shaken' his earlier views. The enhanced role of overland

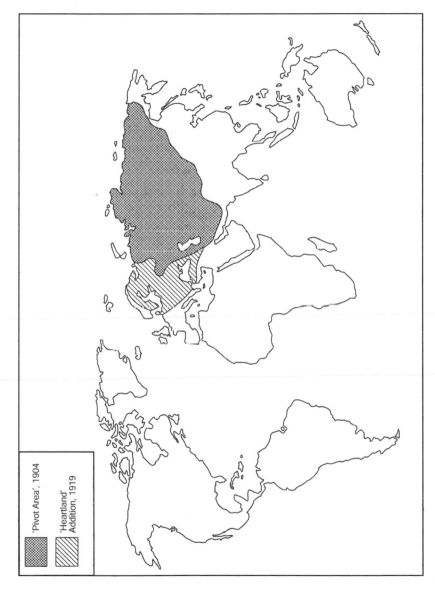

'Pivot Area', 1904

'Heartland'
Addition, 1919

Figure 2.8 Mackinder's heartlands, 1904–19. Source: Mackinder (1904, 435); Mackinder (1919, 135).

transportation had been demonstrated by Germany's ability to fight a war on two fronts, and the eclipse of sea power had been revealed by Britain's failure to exploit its naval superiority. But the Bolshevik take-over in Russia and the likely changes to the political landscape of east-central Europe forced Mackinder to rethink the details of his theory. Although the war had raged around the world and reached its peak of savage intensity in northern France and Belgium, it had really been a struggle for control of eastern Europe, Mackinder claimed, for this was the gateway to the Asian land mass, a region he now dubbed the 'Heartland'. Control of the 'Heartland' was still the key to world power. However, the western limits of this sprawling area now 'intruded' into the core of 'old' Europe (Figure 2.8).

The idea of an expansive Europe stretching from the Atlantic to the Russian steppelands, what Mackinder called 'the real Europe', no longer existed. This had previously been limited by a line running east from St Petersburg to Kazan and then south along the valleys of the Volga and Don to the Black Sea. This was 'the Europe of the European peoples, the Europe which, with its overseas Colonies, is Christendom'. But the 'real Europe', for so long 'a perfectly definite social conception', had been gradually dividing into two divergent and antagonistic regions, east and west. The zone-of-transition between these two areas ran from the shores of the eastern Baltic to the Gulf of Venice, a remarkable prediction of the Cold War division after in 1945 (Figure 2.9). The regions on either side were environmentally, culturally and geopolitically opposed; two different worlds. To the west, in the 'Coastland', a balance now existed between Britain and France, the great powers which had preserved the heritage of 'real' European civilisation. To the east, on the western fringes of the 'Heartland', lay Russia and much of Germany, the most likely Eurasian powers. Russia's destiny as an Asian power was sealed, but Germany, lying athwart this great divide, was the more problematic and unpredictable power; it looked in two directions at once. The division of Europe meant that Germany could no longer claim full European credentials and it seemed likely to turn its back on the Atlantic world of the 'Coastland'. This would spell the end of 'real Europe'; the collapse of European civilisation.

The threat of a future alliance between Germany and Russia was still the ultimate danger, particularly if Bolshevism were to take root in Germany.[34] Equally worrying was the prospect of a German conquest of the Slavic world. This was unlikely in the immediate future but could not be ruled out in the long term. The encircling strategy of the pre-1914 era had failed to prevent Germany waging a war of expansion and had been in any case 'an incongrous . . . alliance between Democracy and Despotism'.[35] An alliance between democracy and Bolshevism would be even more incongruous, Mackinder implied. Germany's future expansion into the 'Heartland' was thus a distinct possibility. The pulse of the 'Heartland' might eventually beat to rhythms laid down in Berlin rather than Moscow or Petrograd.

The lessons were clear and unaltered by the war: 'West Europe, both

Figure 2.9 The end of Europe? Mackinder's divide, 1919. Source: Mackinder (1919, 154).

insular and peninsular, must necessarily be opposed to whatever Power attempts to organise the resources of East Europe and the Heartland'.[36] A few pages later, Mackinder coined his famous dictum to sum up his geopolitical vision:

> Who rules East Europe commands the heartland;
> Who rules the Heartland commands the World-Island;
> Who rules the World-Island commands the World.[37]

Preventing an alliance between Germany and Russia in the east, within the 'Heartland', should be the basis of Allied policy.[38] The only way to achieve this in 1919 was to establish a tier of east European 'buffer states', supported by the western powers, which could separate Germany from Russia (Figure 2.10).[39] Mackinder's vision of a Europe divided into two, mutually exclusive realms reiterated, in the language of twentieth-century geopolitics,

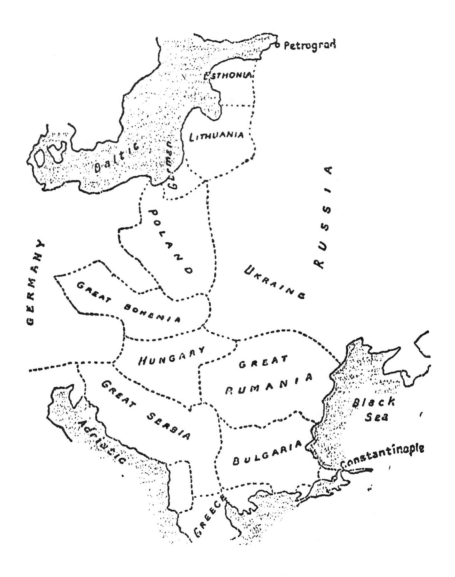

Figure 2.10 Preserving 'real' Europe: Mackinder's buffers, 1919. Source: Mackinder (1919, 207).

the traditional east–west conceptualisation of the continent described in Chapter 1.

Mackinder's 1919 analysis was intended as a lesson in geopolitical realism, a counter-blast to the 'naive idealism' he detected in the proclamations of the US President, Woodrow Wilson, and amongst the advocates of a League of Nations (see Chapter 3). If such an organisation was to be a force for good, it must avoid unworkable concepts such as 'internationalism' or 'one-worldism'. Nor should its agenda be set by lawyers concerned with abstract legal principles. Mackinder's conclusion was that naive 'democratic ideals' ran counter to harsh geographical realities. Appeals to 'universalism' or 'the equality of nations' were futile in a world where access to space and resources was inherently unequal. He claimed:

> There is no such thing as equality of opportunity for nations. Geography, the grouping of lands and seas, and of fertility and natural pathways, is such as to lend itself to the growth of empires, and in the end to a single World Empire. If we are to recognise our ideal of a League of Nations which shall prevent war in the future, we must recognise these geographical features and take steps to counter their influence.[40]

A German alternative: *Mitteleuropa*

Mackinder's predictions generated widespread interest. Not everyone was impressed by his reasoning. George Chisholm, Professor of Geography at Edinburgh University, dismissed Mackinder's famous dictum about Eastern Europe, the 'Heartland' and the World Island as 'a somewhat syllogistic chain of affirmation', and the US geographer, Charles Redway Dryer, criticised his failure to consider the future role of non-European powers, particularly the USA.[41] Leo Amery (then a journalist on *The Times* and subsequently a prominent Conservative politician) argued in response to Mackinder's 1904 paper that a new era of land power and intercontinental railways was unlikely to materialise because another transport revolution was about to occur. The prospect of a world bound together by aeroplanes and global telegraph systems made the geopolitical distinction which Mackinder drew between land and sea power redundant.[42]

Others were attracted by Mackinder's underlining logic, if not his conclusions. The US geographer Frederick Teggart transformed Mackinder's analysis into a quasi-racist denunciation of all things Russian and a spirited defence of Germany as a 'natural' ally of the 'west' in its age-old struggle against the 'east'. Germany, Teggart asserted, was Europe's bulwark against the westward expansion of Russian despotism, now manifested in the new guise of Bolshevism:

> in the opposition to the power of the Heartland, the interests of the Western powers (England, France and Germany) were identical. In

actively opposing the Russian menace to the marginal lands, Germany and Austria were carrying out, for good or ill, the historic mission which Western Europe had entrusted them. Germany is an integral part – the defensive frontier – of Western Europe. If her politico-military organization is a menace, it is the logical product of Western ideas.

A 'United States of Europe', including a fully integrated Germany, was the only solution to 'the new menace of Russia', claimed Teggart.[43]

Such views were common currency in Germany, where Mackinder's ideas were widely discussed. Max Scheler also claimed that the war was a struggle between Europe and Asia, a spiritual crusade which would ultimately ennoble and reinvigorate the European life-force. Unfortunately, matters had become unnecessarily complicated, Scheler argued, by the unnatural alliance which Britain and France, Germany's natural European allies, had forged with alien, Asiatic Russia. Germany had been obliged to carry the sole burden of European culture in this historic struggle, whilst suffering murderous attacks by those peoples who shared a common European interest.[44]

Mackinder's belief that cohesive land empires were likely to replace unsustainable maritime empires scattered haphazardly across the globe also struck a chord in Germany where an emerging school of 'pan-regionalism' likewise predicted that existing European states would be unviable in a future world of rapid, long-distance transport. Europe's political geography was destined to change as its states coalesced to form larger, land-based economic and political units – self-sufficient regional confederations. But the suggestion that a new global configuration would split Europe in two was firmly rejected by most German critics. This was scarcely surprising, of course, for this scenario directly challenged both the territorial integrity of a united Germany and the European credentials of the German people.

A different perspective, based on the idea of *Mitteleuropa* (central Europe), was advanced as an alternative vision of a future Germany in a new Europe. Unlike Mackinder's 'Heartland', an emerging *Mitteleuropa* would preserve Europe's (and Germany's) historic integrity, or so its proponents claimed. The idea that Europe could be divided into three 'natural' regions ranged from west to east, with a central belt, dates back to the mid-nineteenth century.[45] The debate about *Mitteleuropa* and its variants (*Zentraleuropa* or *Zwischeneuropa*) was immensely complex. There was never a clear consensus about its geographical limits or its precise meaning and function. Initially, the concept was understood in economic terms. *Mitteleuropa* was invoked as a new economic space with low 'internal' and high 'external' tariffs which would protect the agricultural and industrial sectors of central Europe and overcome the region's severe economic dislocations and imbalances. This was one manifestation of the late nineteenth-century belief in autarky and protectionism. The economist Friedrich List, whose faith in the integrative capacity of railways was mentioned in Chapter 1, espoused a general central European customs union, building on the German *Zollverein*, during the 1830s and 1840s. A similar idea was advanced by the Austrian Minister of Commerce, Karl von Bruck, through the late 1840s and 1850s. Von Bruck proposed a series of interlocking confederations – German, Austrian, Swiss and Italian – which would act as a counterbalance to the imperial economies to the west (Britain and France) and the east (Russia). The largely economic nature of the *Mitteleuropa* debate during the 1850s is underlined by Robin Okey's analysis of 79 German atlases on sale in 1856. Only four showed the political geography of an area identified as *Mitteleuropa* whereas 24 included thematic economic maps organised on this basis.[46]

Proposals for a customs union embracing Germany, Austria–Hungary and other south central European states resurfaced on several occasions through to the early twentieth century, inspired in part by the establishment of the Pan-American League which suggested (misleadingly) that an economic 'pan-region' was about to emerge, uniting the North and South Americas. An important figure in the campaign for a German economic 'pan-region' was the economist Julius Wolff, whose *Mitteleuropäische Wirtschaftsverein* was an important focus for this debate.[47]

Lurking behind these economic arguments were ideological motives. These emerged more clearly towards the end of the nineteenth century. The idea of a political *Mitteleuropa* won some support on the left. A democratic federal system in central Europe seemed an imaginative solution to the region's brooding ethnic tensions, a radical alternative to the moribund and inflexible autocracy of an Austro–Hungarian Empire in which under half the population were Austrian or Hungarian and where the bulk of the Emperor's subjects, some 47 per cent, were Slavs.[48] Here, ethnic communities, languages and religions intermingled in an unmappable mosaic. The creation of ethnically homogenous microstates seemed hopelessly impractical. Strategically vulnerable and economically unviable, such states would only exacerbate ethnic conflict. A new democratic confederation, on the other hand, in which all communities would have appropriate representation, offered an attractive and progressive alternative to the sclerosis of the existing geopolitical structure and the anarchy of an ethnic free-for-all. A federal *Mitteleuropa*, defined by its ethnic diversity rather than its geographical unity, offered a model from which a wider, federal Europe might emerge.

The German Marxist, Karl Renner, was one of the most articulate advocates of a federal *Mitteleuropa*. His scheme was based on a radical, and highly imaginative, rethinking of how states, sovereignty and political territoriality might operate and has some affinities with the modern notion of 'subsidiarity' (see Chapter 4). In place of discrete, bounded states, each developing their own economic and social policies, Renner advocated the functional separation of responsibilities within a federal system. Wider

economic and political policies could be decided at the federal level by a parliament of elected representatives from the existing states but social questions and matters relating to ethnic or religious groups, including education,
could be handled by a parallel organisation comprising representatives of
each, geographically scattered community.[49] The notion of a socialist central
Europe, with Germany at its core, was extended still further by a few radical thinkers who dreamed of an even larger socialist German empire which
could challenge the capitalist empires of the British and French.[50] The concept of *Mitteleuropa* as a cosmopolitan core within a new, socialist Europe
was by no means an orthodoxy in *fin-de-siècle* Germany, though the idea
that central Europe could be identified by its very diversity was a common
motif in political and academic debate. The geographer Joseph Partsch, for
example, produced an important survey of the region's physical and human
geography based on this theme.[51]

The idea of a political *Mitteleuropa* was also appealing to conservative
thinkers. Theodor Schieman, friend of the ultra-nationalist German Kaiser,
Wilhelm II, listed the dominant objectives which right-thinking German pat-
riots should insist upon at the *fin-de-siècle*: 'a central European customs and
economic union, a settlement of the colonial question on generous lines, the
humiliation of England, the preservation of peace with our allies
Austria–Hungary and Italy and the containment of the powerful Russian
influence'.[52] For Schieman, *Mitteleuropa*, based initially on a customs union,
implied a central European confederation dominated by a greater Germany in
which all German-speaking peoples, 10 million of whom lived outside
Germany in Austria–Hungary and numerous *irridenta* elsewhere, would
finally be united. This was not a *Mitteleuropa* in which ethnic diversity would
be celebrated or even tolerated. The 'smaller peoples' of the region would simply have to accept their gradual demise. Their destiny was to be assimilated
into the dominant German *Kultur*. The call for a German *Mitteleuropa* featured as one of Bethmann-Hollweg's war aims in September 1914 (see above).

> We must create a central European economic association through
> common customs treaties, to include France, Belgium, Holland,
> Denmark, Austria–Hungary, Poland, and perhaps Italy, Sweden and
> Norway. This association will not have any common constitutional
> supreme authority and all its members will be formally equal, but in
> practice it will be under German leadership and must stabilise
> Germany's economic dominance over *Mitteleuropa*.[53]

Mitteleuropa was a complex and ambiguous concept which was invoked
by different constituencies for mutually exclusive reasons. These contradictions came to the fore during World War 1 when the idea of *Mitteleuropa*
was widely debated as one alternative to the existing European geopolitical
structure. Albrecht Penck wrote of a new central Europe (*Zwisheneuropa*)
under German patronage as a means of securing peace and a pathway
towards a united Europe (Figure 2.11).[54] A similar argument was put

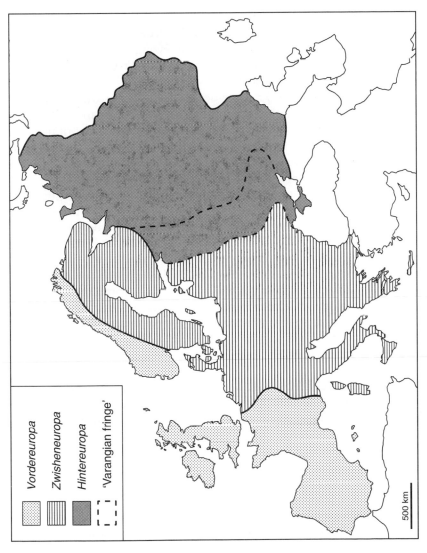

Figure 2.11 *Mitteleuropa*: Albrecht Penck's Europe, 1915. Source: afte Penck (1915, 16); Sinnhuber (1954, 29).

Legend:
Vordereuropa
Zwisheneuropa
Hintereuropa
'Varangian fringe'

500 km

forward the following year by Friedrich Naumann, a former Lutheran pastor and member of the *Reichstag*.[55] Naumann saw a self-sufficient central Europe, dominated by a benign Germany, as the only solution to Europe's manifest geopolitical problems and the chronic incompatibility between German *Kultur*, Anglo–French *civilisation* and the 'Oriental despotism' of Russia. Initially this region would need to be physically isolated from its enemies to east and west by two great 'Chinese walls'. In the long-term, however, *Mitteleuropa* would spawn a wider federal Europe. A united Europe at peace with itself would eventually emerge from its own core region. The essential characteristics of central Europe were the essential characteristics of Europe as a whole.

Naumann's analysis opened a lively debate across the wartime division. His underlying belief in the incompatibility between German *Kultur* and Anglo–French *civilisation* was accepted by several critics in Britain and France. Henri Bergson interpreted the war in precisely these terms, as a clash between two fundamentally different world-views, the products of mutually exclusive national cultures. Peace required either the total eradication of one side or the complete separation of both.[56] Some British critics offered a cautious welcome to Naumann's scheme. A German *Mitteleuropa* sealed off from the rest of the world would spell the end of the 'old' Europe for the foreseeable future, but if that was the price of peace it was perhaps worth paying.[57] The more common Allied response to Naumann's *Mitteleuropa*, however, was angry rejection.[58] Most French commentators dismissed the idea as a flagrant piece of geopolitical propaganda, a thinly disguised justification for German imperial domination. Central Europe was a myth, it was claimed, an invented region of diverse, mutually antagonistic peoples whose 'unity' could only be maintained by force.

Empire and the east: Germany's *Lebensraum*

Paradoxically, the more aggressive proponents of German imperialism were also unconvinced by Naumann's scheme, on the grounds that it was insufficiently ambitious and too conciliatory towards the smaller nations of central Europe. According to these critics, who included key military figures such as generals Ludendorff and Hindenberg, *Mitteleuropa* might serve as a means to an end, a stepping stone towards a greater German empire, but it could never be an end in itself. Naumann's *Mitteleuropa* would place 'unnatural' limits on Germany's potential growth, trapping the nation forever in a central European zone.

These criticisms reflect the increasing scale of German geopolitical ambition after 1890, a trend which can be detected in many different media, including the pulp fiction of the period. Whereas British adventure novels of the 1890s and 1900s were defensive in tone and preoccupied by invasion schemes and sinister enemy spies plotting the nation's downfall, German stories displayed a soaring and aggressive confidence.[59] Karl Kaerger's futuristic novel, *Germania triumphans*, published in 1895, was a revealing example. This foretold an early twentieth-century near future in which an embattled Germany was struggling, in the name of Europe, to break the stranglehold of three enemy 'pan-regions' dominated by the USA, Britain and Russia. War erupts with 'perfidious' France in 1903 from which Germany emerges victorious. A new *Mitteleuropa* alliance of Germany, Italy, Austria and Turkey then destroys Russia to establish a united Europe, under German hegemony. German Europe subsequently does battle in 1912 with the pan-American realm to claim its 'rightful' stake in Latin America. Finally, in 1913, the British Empire is brought to its knees and an ascendant Germany realises the cherished goal of a central African empire, a German *Mittlelafrika*.[60]

This sort of literature reflected the more serious ambitions of the German imperialist lobby, both with respect to Europe and the wider world. Like their counterparts in Britain, France and Russia, German imperialists drew on a quasi-scientific rhetoric. The German concept of *Lebensraum*, or 'living space', was of central importance here. This idea was borrowed from mid-nineteenth-century biology, having been coined by the German biologist Oscar Penschel in an 1860 review of Charles Darwin's *The Origin of Species* (1859).[61] Penschel used the term as an imprecise German alternative to the English 'habitat' or the French 'milieu'. Like many concepts developed in the natural sciences, however, the word soon began to feature in German geopolitical writings, part of a wider process by which a simplified Darwinism was used to explain and predict human behaviour.[62]

The writings of the German geographer Friedrich Ratzel were especially important in this process.[63] Ratzel had trained in the natural sciences and was influenced by Darwinian metaphors and ideas. The main weakness in Darwin's thesis, Ratzel argued, was its failure to consider the importance of space. According to Ratzel, the 'Darwinian' idea of the struggle for existence (*Kampf ums Dasein*) could be equated with a struggle for space. All life forms on the planet were involved in this ceaseless quest for *Lebensraum*. The application of this idea to European geopolitics implied a biological theory of state formation and development, a concept pioneered by Ratzel but further developed by Rudolf Kjellén, originator of the term 'geopolitics'.[64] Both Ratzel and Kjellén saw the nation-state as a natural organism, greater than the sum of the individuals, communities and classes of which it consisted. The state was a living geopolitical force rooted in, and moulded by, its soil. It was an organic entity, the physical embodiment of the popular will and the product of a centuries-old interaction between a people and their natural environment: *ein Stück Boden und ein Stück Menscheit*.[65] Kjellén, in particular, sought to challenge the narrow legal and institutional definition of the state in his path-breaking analysis, *Der Staat als Lebensform*, originally published in Swedish in 1916 and probably the conduit through which the term *Geopolitik* entered Germany. The legal concept of state was, claimed Kjellén, derived from bourgeois republicanism and was therefore rooted in the ultra-rationalism of the French Enlightenment and Revolution. But the state should be far more than this; it should be the vehicle for all the hopes, dreams and aspirations of a people, a measure of their demographic vitality (*Demopolitik*), their economic potential (*Wirstschaftspolitik*), their social vigour (*Soziopolitik*) and their power (*Herrschaftspolitik*).[66] 'All that a man is', wrote Kjellén in 1916, 'he owes to the State. . . . This is where his spirit resides. His valour, his spiritual reality would have no meaning without the State. . . . The nation, and not the individual, is the true hero of history'.[67]

For Ratzel, possession of space could be distinguished from desire for space. The existence of large, vulnerable states (such as the Ottoman and Russian Empires) demonstrated that power did not necessarily derive from the mere possession of space. If this were the case, large states would always grow at the expense of smaller states by virtue of their initial territorial advantage. Strong states, claimed Ratzel, were those which manifested a demographic, economic and cultural capacity which was greater than their existing territorial limits. Such states would inevitably develop expansive tendencies. A state's power was therefore determined by its territorial ambitions rather than by its spatial extent. The acquisition of 'living space', although an obvious manifestation of initial strength, often diminished the lust for territorial expansion and thus undermined a state's dynamism and energy, making it vulnerable to the ambitions of rival states. Ratzel's geopolitics therefore implied a ceaseless process of growth and decline, a permanent waxing and waning in the size of rival states. This was an inevitable consequence of human life, claimed Ratzel, the engine of progress. Ratzel's theory did not preclude the possibility of achieving balance between equally dynamic and vigilant states. Such a balance could only be sustained, however, by the permanent and general desire for greater *Lebensraum*. The endless struggle for 'living space' was thus a fundamental and inalienable geopolitical law.[68]

Ratzel's ideas contain several conceptual confusions. He combined a characteristic German romanticism (derived from Hegel, the *Naturphilosophie* of Heackel and the geographical writings of Carl Ritter which posited an essential life-force determining the nature and actions of all living things) with a Social Darwinism (which relied on fundamentally anti-Hegelian, materialist forms of explanation). Ratzel was thus able to imply an inherent human spirituality operating independently of external forces in some instances and on other occasions to resort to crude environmental determinism. These intellectual contradictions were mirrored by Ratzel's ideological confusion. His romanticism manifested itself in profoundly anti-modern beliefs. Urban–industrial life represented, he believed, a spiritual crisis. Modern cities were soulless places dominated by greed, cynicism and pernicious individualism. Industrial labour crushed the human spirit and destroyed social cohesion by breaking the necessary dialectic between the soil and the people. A rural existence, he implied, in which human beings live in harmony with the rhythms of the natural world, was the only legitimate form of human existence. Yet his geopolitics depended on the very industrial prowess he claimed to abominate. He was, for example, a passionate advocate of an overseas German empire policed by a hugely expanded navy which could rival that of the British.[69]

Ratzel's *Lebensraum* was a theoretical rather than a geographically specific concept.[70] However, the idea could obviously be applied to legitimate German territorial ambitions, particularly in the east. The ancient Germanic myth of the *Drang nach Osten*, the drive of the Teutonic knights to the east, was thus reborn in the 'scientific' language of Ratzelian geopolitics. Germany's historic destiny was deemed to lie not merely in central Europe, its immediate arena, but in eastern Europe and beyond. This eastern realm, currently part of the Slavic world, was where a German *Lebensraum* would emerge. Here was the space for a new German empire in the coming century; the zone where the burgeoning German population could begin their new era of colonisation.

The Russian realm: pan-Slavism

Germany's territorial ambitions in the east were both a reason for, and a response to, the British, French and Russian attempts to establish an encircling system of alliances (see above). The eventual creation of this system was a triumph for pro-western elements in Russia's political élite, the inheritors of the Petrine view of Russia as a western, European power (see Chapter 1). But this was not the only vision of Russia's potential role. The preceding decades of suspicion and antagonism between Russia and the western powers had been sustained by a romantic pan-Slavic nationalism which spawned its own visionary geopolitics in the decades after the Crimean War. Russia's Slavophiles rejected the materialism and commercialism of western Europe and insisted that the Slavic arena, so often cut in two by an east–west cultural division of Europe, was a distinctive spiritual world, belonging neither to the 'west' nor to the 'east'. Russia's position as the leading Slavic nation whose territory lay athwart this outdated cultural divide gave the nation a special historic mission to forge a new, Eurasian geopolitics transcending the old east–west divisions.[71]

Pan-Slavism was a crusade, led by Russian intellectuals, on behalf of all Orthodox and Slavic peoples, particularly those who lived under Ottoman and Austro–Hungarian 'oppression'. Although rarely a dominant influence on Russian foreign policy, pan-Slavic arguments had considerable emotional appeal which occasionally shaped official responses and attitudes. The Serb and Bulgarian revolts against Ottoman rule in the mid-1870s were loudly supported in pan-Slavic circles and the Russian government gave tacit, and some active, support to Russian soldiers who chose to fight in the

Serbian army. Intellectuals such as the novelist Fedor Dostoevskii believed that war with the Ottoman Empire was a spiritual necessity.[72]

The pan-Slavic doctrine, sometimes expressed as 'one race, one language, one religion for all Slavs', looked forward to a new empire whose capital would be located not in St Petersburg or Moscow but in a reconquered Constantinople, or 'Tsar-Grad'. Here, the pan-Slavs claimed, the 'cross would once more be seen upon the dome of St Sophia'.[73] The startling geopolitical ambitions of this movement were made clear in an extraordinary poem from 1849 by Fedor Tiutschev entitled 'Russian Geography':

> Moscow and Peter's city and the city of the Constantines –
> These are the secret capitals of Russia's realm . . .
> But where are her bounds and where her frontiers?
> To north and east and south and towards the evening light?
> Fate will reveal them to coming generations.
> Seven internal seas and seven great rivers . . .
> From Nile to Neva, from the Elbe to China,
> From Volga to Euphrates, from the Ganges to the Danube . . .
> That is the Russian realm. . . .[74]

One of the most influential Slavophiles was Nikolai Danilevskii whose *Russia and Europe* (1869) provided a 'breathtaking spatial vision of a future Russian-dominated Pan-Slav Union stretching from the Adriatic to the Pacific'.[75] In Danilevskii's view, western Europe, which had arisen from the classical civilisations of the Mediterranean, had now sunk into a spiritual morass of materialism and individualism. A war between the Romano–Germanic civilisation of western Europe and the Greko–Slav civilisation of Eurasia was inevitable. From this, a renewed Slavic–Orthodox civilisation would emerge triumphant to provide a fresh moral and spiritual order based on a common Orthodox religion and a single Slavic language.[76]

A central motif in the pan-Slavic geopolitical imagination was the idea of an advancing Asian frontier of Slavic civilisation, a redemptive zone of expansion where the spirit and values of the emerging empire would be forged. The historians Mikhail Pogodin and Vasiliy Klyuchevsky saw Russia's past and future in these terms, as did many of the leading members of the Imperial Geographical Society such as V.V. Grigorev, N.N. Muravyov, G.I. Nevelskoy, Pyotr Semyonov and Mikhail Venyukov whose explorations and writings on Asia perpetuated the idea of Russia's 'national mission' in the east.[77] The most famous account of Russia's Asian destiny was provided by Sergei Solovyov's multi-volume *Istoria Rossii* (1851–79).[78] Similar views were also articulated by Dostoevskii in 1881: 'To us, Asia is like the undiscovered America. . . . With our aspiration for Asia, our spirit and forces will be regenerated'.[79] By 1906, Dimitry Mendeleev was arguing that the Russian 'rejection' of the west in favour of the east was beginning to manifest itself in a spontaneous eastward movement of the Russian people, the first manifestation of an emerging Eurasian world. Despite Russia's wartime alliance with France and Britain, pan-Slavic propagandists such as Veniamin Semyonov-Tyan-Shansky insisted (like Scheler in the 'west') that the conflict was really a struggle between west and east. Russia's historic destiny was to lead a spiritual renaissance of the Slavic east which could be restructured after the war into a colonial federation of 19 self-governing 'dominions'.[80]

Belief in the spiritually redemptive qualities of Asia was surprisingly widespread in western European countries as well. The German idea of an eastern *Lebensraum* was often expressed in this rather mystical register. Some German expansionists wrote of the mountains of Tibet and Central Asia, poised between the ancient civilisations of China and India, as the wellspring of 'Aryan' cosmography, 'the navel of the earth'.[81] British explorers and mountaineers in the region were also inspired by such beliefs.[82] Lord Curzon, the post-war British Foreign Secretary and a passionate geographer who claimed a special expertise in this part of the world, wrote in 1898 of the four great rivers of Asia which 'rocked the cradle of our race'.[83] The same theme informed the US geographer Ellsworth Huntington's remarkable 1907 essay, *The Pulse of Asia*.[84]

As Mark Bassin has demonstrated, the pan-Slavic frontier thesis in respect to Asia presents obvious parallels with Frederick Jackson Turner's simultaneous description of America's western frontier. This was also seen as the nation's 'cutting edge' in its 'manifest destiny' to fill the 'empty' space between the Atlantic and the Pacific; it was the defining motif of the US national identity.[85] Such views were also evident in British attitudes to their own 'imperial frontier', a theme developed with typical grandiloquence by Lord Curzon in his Romanes Lecture at the University of Oxford, where he was Chancellor, in 1907.[86] 'Frontier' theories were recognisably part of a *fin-de-siècle* enthusiasm with 'closed-space' thinking, a fashion which Mackinder so eloquently represented.[87] Indeed, the geopolitics of Russian pan-Slavism and Germanic ideas of an eastern *Lebensraum* lent support to Mackinder's vision of a new Eurasian empire rising in the east to replace an 'old Europe' which had been forged out of the Mediterranean and Atlantic civilisations. Although pan-Slavism and *Lebensraum* were in direct competition, they drew on a common belief in a mystical national destiny in the east. Both visions also foretold if not the end of Europe then at least its wholesale transformation and a dramatic eastward shift in its centre of gravity.

Europe and its peoples: racial geopolitics

Most of the ideas so far discussed, from Mackinder's 'Heartland' thesis to Danilevskii's pan-Slavism, were influenced by racial thinking. 'Racialism' (the idea that humanity is divided into distinctive 'races' with associated physical, social and intellectual traits) was a doctrine accepted by most shades of political opinion in late nineteenth-century Europe. Ideas which now appear ludicrous and offensive were common currency. Nineteenth-century racialism was an extension of the 'civilisational' debates of the eighteenth century (see Chapter 1), the dark side of the Enlightenment urge to classify, categorise, order and control. The post-Darwinian debate about 'race' was expressed, however, in the language of the biological sciences, the principal rhetoric of *fin-de-siècle* social and political discourse. Entire scholarly disciplines, notably anthropology, developed to pursue the 'scientific' study of racial characteristics.[88]

'Scientific racism' was, then, a central component in *fin-de-siècle* European geopolitics.[89] A conviction in the racial superiority of white, European peoples over the inhabitants of other regions was a central tenet of European imperialism.[90] By the end of the nineteenth century, racial theories had also begun to influence geopolitical debate about Europe itself. A simple question lay at the heart of these debates: what rôle should racial factors play in defining Europe's external limits and its internal geopolitical structure?

At the risk of oversimplifying, one can identify two perspectives on the European racial question in this period. The first, more reactionary, position interpreted racial characteristics as relatively static and unchanging. This stemmed, in part, from a 'polygenetic' understanding of human history. The existing racial variety was deemed to have emerged from several different original human groups. This rigidly hierarchical view was advanced by the French 'father of racial ideology' Arthur de Gobineau in his *Essai sur l'inégalité des races humaines* (1853–5). Gobineau's principal fear was that the intermingling of 'pure' racial stock, particularly in the burgeoning industrial cities of Europe and North America, was creating a degenerate form of humanity.[91] Racial groups should be kept firmly apart, he argued, to avoid miscegenation and to protect the purity of the 'superior' races. This was, he admitted ruefully, an increasingly difficult task given the ever-rising mobility and urban–industrial development of the modern age.

The second perspective interpreted race in more fluid terms, as a changing set of environmentally determined characteristics. This argument was associated with a 'monogenetic' view of a single original humanity from which the existing spectrum of modern races had emerged as a result of a long process of interaction with an endlessly varying natural environment. Whereas 'polygenetic' racialism downplayed environmental influences, 'monogenetic' racialism placed considerable emphasis on the external world of nature.[92] According to this argument, racial intermingling could, in the correct circumstances, facilitate racial 'improvement' rather than degeneration. Tangible improvements in even the most 'debased' human races could be produced by the careful modification of prevailing environmental conditions coupled with judicious educational strategies. In its more liberal manifestations, this view merged into a utopian 'one-worldism' which foresaw the emergence of a 'universal race', a concept debated at the Universal Races Congress in London in 1911.[93] Eugenics, the idea that a civilised and progressive society had a right, indeed a duty, to intervene in the process of human reproduction to preserve and enhance the quality of the race, sprang from this view of a race as a malleable phenomenon. Eugenics culminated with the horrific racial and genetic experiments of the Nazis (see Chapter 3) but the belief that human breeding should be controlled to eradicate anti-social, immoral and physically undesirable traits was enthusiastically embraced by many liberal and even socialist commentators at the *fin-de-siè-cle*. The intervention to control human sexuality seemed merely a logical extension of a reformist agenda which had already embraced intervention in the economy and social order. Eugenics was part of a wider faith in the power of rational science to 'cure' social and political problems.[94]

These two racial perspectives represented differences in emphasis and tone rather than clearly defined schools of thought. Variations on both can be detected in the main *fin-de-siècle* texts on Europe's racial condition. In his 1899 *Die Grundlagen des Neunzehnten Jahrhunderts* [The Foundations of the Nineteenth Century], the 'Germanised' Englishman, Houston Stewart Chamberlain, wrote of Europe's races as arising largely independently of external environmental influences. Races were the product of inherent, unchanging biological characteristics mixed with cultural and spiritual qualities rooted in a sense of community and belonging. Chamberlain (son-in-law of the anti-semitic composer Richard Wagner and later Adolf Hitler's favourite racial theorist) helped to establish the myth of an 'Aryan super-race' whose destiny it was to dominate Europe.[95] A more flexible, environ-mental reading of Europe's racial geography was provided in William Z. Ripley's *The Races of Europe*.[96] Ripley, a US professor of sociology, wrote of three pure races, the Teutonic (or Nordic), the Alpine and the Mediterranean, which had intermingled to produce Europe's complex racial mosaic. Ripley's racialism was scarcely less spurious than Chamberlain's and was based on a panoply of physiological evidence including phreno-logical indexes, hair colour and skin pigmentation (Figure 2.12).

Geopolitical debate about Europe was directly informed by both these perspectives on 'the racial question'. According to the first argument, certain European races were inherently less developed, civilised and dynamic and therefore possessed fewer rights. In Kjellén's geopolitical analysis, for exam-ple, the failure of some races to develop a modern nation-state was itself evidence of their racial inferiority. The 'smaller' races of Europe merely existed within a given territory; they could never really possess or develop it. Rather than controlling their environment, they were mastered by it. As a result, these people were destined to be overrun by the more civilised, nation-building races whose 'higher' needs propelled them to expand into 'undeveloped' spaces. The task of the 'superior', dynamic races was to main-tain their racial 'purity' whilst expanding into territory occupied by these 'lesser races'. The destiny of the latter was either to decline or remain for-ever under the suzerainty of the more civilised, colonising power. The more uncompromising schemes for an enhanced German *Lebensraum* in the east

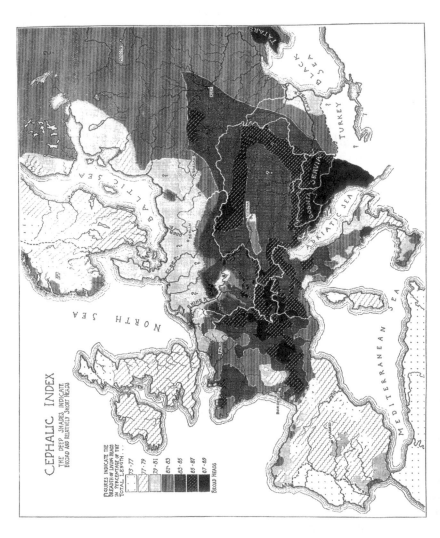

Figure 2.12 Racial geopolitics: William Z. Ripley's Europe, 1899. Source: Ripley (1899, opposite 52).

were, therefore, based on the idea of inimical racial differences. The Slavic peoples in the east were, it was claimed, a 'subculture'; an inconvenient obstacle to the creation of a racially pure German empire. Ironically, the more extreme pan-Slavic imperialists adopted similar attitudes to the 'small peoples' of Asia and Turkey, inhabitants of those regions which were destined to make up a future Slavic empire.[97] The second, more flexible and 'assimilationist' perspective on race also informed geopolitical discussion. Ratzel's vision of a German *Lebensraum*, for example, owed more to this view than to the rigid, 'separationist' ideas derived from Gobineau. For Ratzel, a new German empire would be invigorated and rejuvenated by racial mixing. Many pan-Slavic theorists also spoke of an ethnically and religiously mixed Slavic empire which would ultimately produce new hybrid civilisations.

Differences of opinion on the 'racial question' added further layers of complexity to geopolitical debates. The outbreak of war in 1914 gave a new urgency to this somewhat academic discussion. Initially, the war was itself interpreted in racial, Social Darwinian and eugenic terms. Through 1914 and 1915, many German, British and French commentators wrote of the war as a moral and physical necessity, an uplifting struggle which would enhance the virility and biological prowess of the victorious race, not least by eradicating the 'excess population'.[98] War, claimed the physician Sir Arthur Keith, was 'nature's pruning hook' and ensured racial vigour and regrowth.[99] For the geographer Sir Thomas Holdich, war was 'the first and greatest civilizing agent . . . a fierce struggle for the survival of the fittest . . . the heritage of the world's overgrowth of population'.[100] The unprecedented death tolls of 1916 destroyed this argument, possibly for good, and the latter part of the war produced an alternative (and only marginally less disturbing) view of war as an essentially 'dysgenic' force. Far from strengthening the racial stock, it was claimed, modern warfare tended to weaken it by removing the best and brightest of the young male line. In the words of the German medic G.F. Nicolai, 'the human riff-raff . . . stay at home . . . [to] dress their ulcers while the brave strong young men lie rotting on the battlefield'.[101]

It was the recognition that Europe's political geography would be transformed by the war which led to the most interesting literature on the geopolitical significance of race, ethnicity and language. The idea that nations could only survive if based on a high degree of racial and/or linguistic coherence became increasingly popular, part of a wider intensification of *fin-de-siècle* debates about what constituted a nation. Ernest Renan's famous 1882 lecture at the Sorbonne, *Qu'est-ce qu'une nation?*, had started this debate.[102] Although Renan showed little interest in race, others were attracted by the superficial simplicity of race as an index of national belonging. Prior to 1914, the idea of 'national self-determination' for all racial groups had received only limited support, mainly from the left. This was not surprising for national self-determination directly challenged the integrity of all the multi-ethnic European imperial states, on both sides of the wartime divide. The orthodox view, equally common in Germany, France, Britain and Russia, was that only racial groups of sufficient civilisation and demographic weight were capable of sustaining a modern nation-state. The smaller, 'less developed' peoples of Europe were destined to live within larger geopolitical structures dominated by more powerful ethnic groups. During the war, however, self-determination became more popular in Allied countries, though the idea was invoked in a partisan and selective fashion. Unlike Allied empires whose territorial legitimacy stemmed from the consent of their various constituencies, the Central Powers were deemed to be racially unsustainable and maintained only by force. They therefore had no place in a new European political geography.

The racial critique of the Central Powers focused especially on Austria–Hungary and the Ottoman Empire but was also applied to Germany. According to Ernest Barker, the racial cohesion of the German Empire was more apparent than real. There were, he insisted, dozens of 'submerged nationalities' in Germany whose distinctive cultures and languages had been brutally suppressed.[103] The real enemy, not only of the Allies but also of Germany's oppressed minorities, was Prussia, a region which had produced a soulless, robotic people, a 'machine-culture' obsessed with violence and war.[104] Wartime pleas for new, racially-cohesive states on former German or Austro–Hungarian lands received a ready audience in London, Paris and New York.[105] Meanwhile, wartime propaganda maps and accompanying pamphlets offered numerous alternative visions of a reconstituted post-war central Europe in which the old empires would be dismantled (Figure 2.13).[106]

Germany's wartime propaganda countered with uncompromising assertions about the unity of purpose of all the races, religions and cultures fighting against the Allies. Some of the most revealing examples from a vast literature were the many illustrated pamphlets edited by the Heidelberg geographer Alfred Hettner entitled *Der Deutsche Krieg*. These were produced by a group of academics and businessmen, closely linked to Bethmann-Hollweg, which included Penck as well as both Max and Alfred Weber.[107] According to Hettner:

> The English regard themselves as a chosen people and believe themselves entitled to rule the seas and the world; but we deny them this right and make ourselves their equals. The Russians claim a right to extend their empires even further, but we challenge their right, too, with our own. . . . We want to be the educators of the world, to carry our culture out into the world. The German Ideal shall heal the world. Germany . . . cannot live from the crumbs which fall from the World Powers' table; she must be a World Power herself, the peer of the others.[108]

Hettner offered a series of proposals for a new world order which would, he claimed, be just to all sides in the conflict.[109]

Figure 2.13 'La paix draconienne': a French view of a future Europe, *c.* 1917. Source: Cambridge University Map Library, Maps. 23.91.94. Note: the text accompanying this extraordinary map, whose title in English reads 'The future Europe: what the Allies must impose to ensure a definitive peace in Europe', refers to dismembering the German and Austro–Hungarian Empires, destroying Prussian influence and reconstituting Poland. Here Germany is divided into six independent states – Hannover, Westphalia, Saxony, Bavaria, Württemberg and Prussia. A great 'buffer zone', 100 kilometres wide, stretches from the Dutch to the Swiss borders. France's eastern frontier is extended to the west bank of the Rhine almost as far north as Coblenz.

Racial interpretations of Europe's history and present condition were especially popular in wartime USA. Race had long been a central preoccupation of American social commentators as the nation's population was the product of massive, multiracial immigration from Europe. Ripley's work provided an early benchmark which was extended by Madison Grant's sweeping historical survey of Europe's eclipse, *The Passing of the Great Race; Or, The Racial Basis of European History.*[110]

Pro-war Americans saw it as a patriotic duty to warn their fellow citizens about the calamitous implications of a German victory. Douglas W. Johnson, a leading geomorphologist from Columbia University, was a vocal opponent of Prussia's militarism which, like Ernest Barker, he interpreted as an inherent racial characteristic.[111] Johnson was Chair of the American Rights League which sought to mobilise US public opinion behind the Allied cause. Immediately after his government declared war, Johnson toured the nation's city halls lecturing on 'The Peril of Prussianism', an emotive appeal which was subsequently the basis of an early propaganda film. Johnson's lecture was illustrated by a series of startling maps showing the apparently irresistible rise of Prussia as a black 'plague' spreading across the face of Europe (Figure 2.14).[112] The 'American ideal', claimed Johnson, in which 'the government is the servant of its citizens and exists for their benefit', is fundamentally at odds with the 'the Prussian ideal' in which 'the individual is the servant of the government and exists only for the benefit of the government'. A Prussian-dominated Europe would compromise the very essence of what it meant to be American:

> America has decided that splendid isolation is no longer possible in a world rendered wondrous small by the swift steamship and express train, the telegraph and telephone, the cable and wireless telegraphy. We *are* our brother's keeper, and have entered on a policy of international co-operation, first to compel peace, and then to preserve it undisturbed from future assaults by autocratic militarism. We have pledged our faith and must fight to make the world safe for democracy.[113]

The idea that a post-war Europe might learn lessons on race and geopolitics from the US was attractive to many leading US politicians and academics, including President Woodrow Wilson. Wilson had an idealistic, some would say naive, faith in the power of reason, logic and objective facts to solve even the most intractable problems. He genuinely believed that the US could adjudicate on Europe's geopolitical crises as an expert but neutral observer. A few months after the US entered the war, Wilson established an ambitious academic project, the so-called House Inquiry, to gather together in a single archive all relevant information – historical, ethnic, cultural, linguistic, religious – which would be required to establish a new, rational political geography for post-war Europe.[114] The House Inquiry, named after its chairman Colonel Edward Mandell House, was to prepare the ground for a future peace conference and secure a privileged position for the US which would represent the higher ideals of peace, justice and honour. The Inquiry was headquartered at the New York offices of the American Geographical Society whose indefatigable Director, Isaiah Bowman, became 'the presiding genius of the organization', though he was technically only second-in-command as the Chief Territorial Specialist.[115] Under Bowman's influence, the House Inquiry became the most exhaustive exercise in geographical and historical data collection ever attempted. Over 150 leading US scholars were involved, representing virtually every discipline. Geography and cartography were the bed-rock concerns of the Inquiry, a

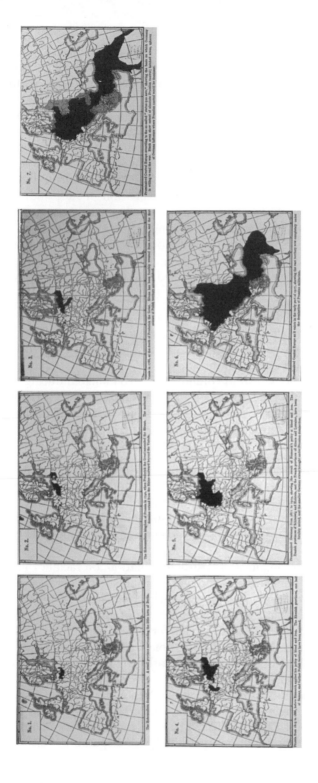

Figure 2.14 'The Peril of Prussianism': Douglas Johnson's Germany, 1917. Source: Johnson, D.W. (1917b, opposite pp. 2, 10, 18, 26, 34, 42 and 50). Note: these maps depict Prussia's supposed territorial and geopolitical influence in 1477, 1740, 1786, 1815–1866, 1871–1914 and 1917. The final image suggests the likely extent of Prussian influence if the war were to end with a negotiated peace rather than a clear victory for the Allies.

fact underscored many years later by Charles Seymour, an historian who co-ordinated the US work on Austria–Hungary and subsequently became President of Yale: 'The sense of the relations of geography to all fields pertinent to peace', wrote Seymour, 'spread contagiously among economists, historians and lawyers'.[116]

The House Inquiry produced no single report (though its members wrote hundreds of newspaper and journal articles). It did, however, seek to legitimate the Wilsonian ideal of 'national self-determination' based on ethnicity. Precisely how ethnicity could be measured was a thorny problem, of course, though many members of the Inquiry became convinced that language represented the best 'mappable' approximation of race and, therefore, of nationality. Leon Dominian's *Frontiers of Language and Nationality in Europe*, published in 1917 under Bowman's editorship and with a preface by Madison Grant, was a serious attempt to construct a new political geography based primarily on language (Figure 2.15). Language determines national consciousness, claimed Dominian, because 'words express thought and ideals'.[117]

> [I]t may be submitted that the advance of civilization ... has been marked by the progress of nationality, while nationality has been consolidated by identity of speech. ... Men alone cannot constitute nationality. A nation is the joint product of men and ideas. A heritage of ideals and traditions held in common and accumulated during the centuries becomes, in time, the creation of the land to which it is confined. Language, the medium in which is expressed successful achievement or hardship in common, acquires therefore cementing qualities. It is the bridge between the past and present.[118]

Dominian's linguistic determinism represents an intriguing illustration of the wartime quest for a radical, 'scientific' alternative to the illogical arrangements which had produced such carnage and destruction. Needless to say, his analysis made several heroic assumptions and ignored crucial issues, notably bilingualism, perhaps the defining characteristic of Europe's principal flashpoints.

Although the underlying idea of 'self-determination' on racial or linguistic grounds grew in popularity as the war continued, few believed that it should be applied as a general guiding principle, not least because (as we have seen) the concept challenged the legitimacy of all multi-ethnic states, including those on the Allied side. Partly for this reason, Wilson shrank from using the term in his famous 14-point plan, issued on 8 January 1918 to honour his election pledge 'to make the world safe for democracy' and to inaugurate what we would now call 'the peace process'. Instead he invoked the vague notion of 'autonomous development', though only with respect to 'the peoples of Austria–Hungary' (Point 10) and the 'nationalities' of the Ottoman Empire (Point 12). Even his explicit call for 'an independent Polish State ... [to] include the territories inhabited by indisputably Polish popu-

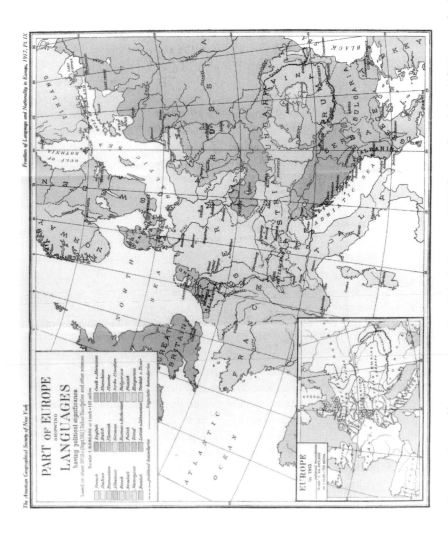

Figure 2.15 The geopolitics of language: Leon Dominian's Europe, 1917. Source: Dominian (1917: opposite 334).

lations' with its own 'free and secure access to the sea' (Point 13) contained no specific geographical details.[119]

European suspicion of US idealism was fuelled by a genuine fear that a fragmented, 'Balkanised' Europe would descend into permanent chaos and endless territorial squabbling. A more pragmatic analysis of Europe's future which sought to accommodate the continent's racial complexity with the need for international co-operation, particularly during the period of post-war reconstruction, emerged from the British weekly review, *The New Europe*, established in 1916 by R.W. Seton-Watson of King's College, London. Ironically, some of the ideas presented in *The New Europe* were derived from pre-war German critics of Austria–Hungary and those, like Karl Renner, who had advocated a liberal or socialist *Mitteleuropa* (see above). The dominant influence on the magazine was Thomas G. Masaryk who settled in London in 1914 to advance the cause of a free and independent Czechoslovakia, the state he was subsequently to lead.[120] *The New Europe* (which included Mackinder on its editorial board) was dedicated to 'the emancipation of the subject races of central and south-eastern Europe from German and Magyar control'. Its articles, many accompanied by garish coloured maps, focused on the immense scale of Germany's geopolitical ambitions and proposed alternatives in which a defeated Germany would be substantially diminished and Austria–Hungary entirely removed from the map (Figure 2.16). Undemocratic and authoritarian, the dual monarchy was deemed to have no place in 'the new Europe'. Simultaneously too large, too weak and too ethnically diverse, it would remain a focus of conflict as long as it survived. But most articles in *The New Europe* rejected the idea of a multitude of small, ethnically 'pure' states. Instead, at least two democratic and multi-ethnic federations were repeatedly proposed, combining Czech and Slovakian peoples in the north, and Serbs, Croats, Slovenes and Macedonians in the south. Seton-Watson and Masaryk hoped (like Renner) that multi-ethnic, federal democracies in the heart of Europe would demonstrate the viability of devolved and decentralised government and would, in time, spawn a wider European federation. These states would also serve an important strategic role: they would be powerful enough to prevent a resurgent Germany expanding to the south.[121] Sir Arthur Evans, a leading archaeologist and a member of the editorial board of *The New Europe*, was a passionate advocate of the new 'South Slav' (Yugoslav) state on these grounds. Germany's ultimate objective, he argued in 1916, was the creation of 'a great State running from Hamburg to Basra, from the North Sea to the Persian Gulf. . . . It is for us to devise a counter plan based on the existence of a South Slav state capable of democratic and progressive culture and [a] life of [its] own'.[122] The most respected and influential prophet of such an arrangement was the leading Serbian geographer Jovan Cvijic who spent the latter part of the war in Paris. Cvijic published two important works in French during this period, the second being his universally praised magnum opus on the human geography of the Balkans; a work which appeared on

(a)

(b)

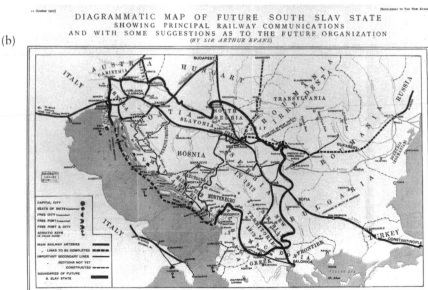

Figure 2.16 'The New Europe': R.W. Seton-Watson's Europe (a) 'Why Germany Wants Peace Now' and (b) 'Diagrammatic Map of Future South Slav State'. Source: *The New Europe* – I, 13 (1917) opposite 420; IV, 52 (1917) opposite 416.

the eve of the Armistice.[123] By the mid-point of the war, race had become the basis of the one-sided 'self-determination' argument, an Allied geopolitical critique of the Central Powers.

The Scottish biogeographer, Marion Newbigin, also contributed to the debate on race and Europe's political geography, particularly in respect to the Balkans.[124] Drawing on theories of species diversity and environmental stability more commonly invoked today, Newbigin developed an environmental critique of the dehumanising and homogenising influence of 'advanced industrialism'. Rampant modernity, which she associated particularly with Germany, was destroying Europe's diversity, she claimed, including its various nationalities and 'traditional' cultures.[125] Geography was the study of regional difference and it was the task of the geographer, she insisted, to preserve and enhance this variety; to challenge the soul-destroying, environmentally harmful and politically unsustainable modernity represented by Germany. The war had been fought in order to achieve this. But the creation of racially 'pure' mini-states, the 'one-race–one-nation' ideal, was no answer, for this led not to real diversity but rather to homogeneity within small, isolated micro-states. Race, she declared, was an historical invention, 'a form of political ideal' which should never be enshrined in the landscape. She argued:

> The physical characters of race have been steadily diminishing in importance during the long slow upward climb to civilization. . . . [T]his blending and blurring of racial characters . . . prove[s] that the physical characters of race are now of minor importance in the struggle for existence. . . . [W]hat makes a nation is not only race – whatever race may mean – not only religion, nor language, nor history, nor tradition . . . but, partially at least, a community of economic interests dependent on geographic factors.[126]

The scientific frontier

The House Inquiry was by no means the only wartime attempt to produce intellectually compelling alternatives to Europe's discredited political geography. Several European governments also 'recruited' their leading scholars to concoct territorial schemes for different parts of the continent.[127] The French government, under its new Prime Minister Aristide Briand, established the *Comité d'Études* in early 1917 to devise a blueprint for Europe's political geography, the basis of France's negotiating position. The *Comité* was set up partly in response to criticism, particularly from the USA, that the war had been prolonged by the failure of European governments to clarify their war aims.[128] The *Comité* drew on the peerless erudition of French academia, particularly historians and geographers. Its Chairman was Ernest Lavisse, France's leading historian, and its Vice-Chairman was Paul Vidal de la Blache, doyen of the French school of regional geography. The two had a long working relationship. Lavisse edited the massive *Histoire de la France*, the 'official' version of the nation's past at the close of the nineteenth century, the opening volume of which was Vidal de la Blache's evocative *Tableau de la géographie de la France.*[127]

Despite their superficial similarities, the House Inquiry and the *Comité d'Études* could not have been more different in their guiding principles and methodologies. From New York came 'New World' solutions, at once radical, ambitious and idealistic; from Paris came 'Old World' proposals which spoke of the fears, anxieties and resentments of the past.[130] Thousands of hours were spent labouring over maps and manuals in the Sorbonne and the *Bibliothèque Nationale*, and dozens of expert witnesses were interviewed from business, industry and commerce. The result was a hefty two-volume publication, appearing immediately after the armistice. The first, much longer, volume was entirely dominated by a single question: where should the Franco–German border be positioned? The *Comité*'s answer, which would have surprised no one, was the Rhine. This was the 'natural' border between Teuton and Gaul, and France had an historically undeniable claim to a presence on the west bank of the river in Alsace. The return to France of the 'lost provinces' of Alsace and Lorraine, ceded to Germany in 1871, was 'indisputable'. A mass of historical, anthropological, linguistic, architectural and geographical evidence was assembled to demonstrate the 'Frenchness' of these twin provinces, including studies of farm buildings and field systems, Gallic landscape features which the decades of Teutonic corruption had failed to efface.[131] A supplementary territorial claim was also made in respect of the frontier coal fields of the Saarland, north of the city of Saarbrucken in the German Rhineland. There was a natural geological and economic unity between the Saarland and the iron ore seams of the Lorraine, claimed the report. Pre-war Germany had developed the resources of both regions together, and new road and rail communications had made this whole area into an integrated economic space. Returning Lorraine to France without the Saarland would harm the economic development of both areas. Other, more or less familiar arguments were also developed including the insistence that Germany pay massive reparations and establish a demilitarised zone in the remainder of the Rhineland.[132] No counter-arguments were possible, it appeared. In the words of the main geologist on the *Comité*: 'We impose our wishes. We impose them completely. I do not consider any other hypothesis as possible, as worthy of consideration'.[133]

The work of the *Comité d'Études* was significant in many respects. It demonstrated the faith which political élites were willing to place in academic expertise. It also revealed that historical and geographical research, even the ostensibly neutral, poetic work of Lavisse and Vidal de la Blache, was directly informed by political objectives.[134] But the most striking feature of the Comité's report was the overwhelming significance of the Franco–German frontier. Although detailed French policies were also worked out for the rest of Europe in the second volume of the report, the fact that so much expertise was expended to assert a single, supposedly 'indisputable' claim reveals a great deal about French, and arguably European, insecurities.

Few European political leaders believed that the continent's new political geography could be decided by reference to abstract guiding principles. National security was the all-important criterion and political borders needed to be constructed, and then justified, to preserve and enhance a nation's defensive capacity. Some of the existing imperial states, such as Austria–Hungary, were clearly untenable and here linguistic or racial considerations had a role to play in devising a new geopolitical order. Elsewhere, however, changes should be made on entirely different grounds; the objective was not ethnic cohesion but the creation of defensible national spaces.

The wartime interest in the political boundary extended the *fin-de-siècle* interest in frontiers of settlement and their role in nation-building. The idea that a science of frontiers and boundaries might be developed was mooted by Curzon in his 1907 Romanes Lecture (see above). Here Curzon distinguished between boundaries (the mappable lines of political demarcation between nation-states) and frontiers (the zones about these lines). 'Frontiers', he claimed 'are . . . the razor's edge on which hang suspended the modern issues of war or peace, of life or death to nations'. Moving from a consideration of 'natural frontiers' (seas, deserts, mountains, rivers, forests and swamps) to 'artificial frontiers' (walls, neutral buffer zones and other 'modern expedients'), Curzon optimistically perceived a 'general tendency' towards a more 'scientific', rational frontier. 'Frontiers', he concluded, 'which have so frequently and recently been the cause of war, are capable of being converted into the instruments and evidences of peace'.[135]

A decade later, Curzon's optimism seemed sorely misplaced. However, his quest for a 'scientific frontier' which could guarantee peace had become the 'Holy Grail' of wartime geopolitics, a topic on which everyone seemed to have an opinion. The angry dispute in Britain between Lionel W. Lyde and Sir Thomas Holdich was illustrative of the differing views. Lyde was Professor of Economic Geography at University College London and the author of a widely read 1913 text on *The Continent of Europe*, a work filled with anthropological and racial theories which were bizarre even by the standards of the period.[136] Holdich was Superintendent of Frontier Surveys in India until 1898 and subsequently a member of the British Arbitration Tribunal which adjudicated on many international boundary disputes in Latin America.[137] He was a powerful figure in the British geographical movement and became President of the Royal Geographical Society in 1917.

Lyde argued that an international agency should take responsibility for designing the political geography of post-war Europe, perhaps involving a grand anthropological survey to determine the racial character of each and every European citizen.[138] Lyde's version of the 'scientific frontier' was based on ethnic and linguistic criteria, though he argued that 'assimilative' cultures (such as France) should be given far more territory than non-assimilative cultures (such as Germany). French *civilisation* was inherently generous and expansive with a 'proven' ability to absorb minority peoples who were either too small in number or too scattered geographically to deserve their own nation-state. German *Kultur*, on the other hand, was rooted in its own Teutonic soil and incapable of incorporating such groups. The war, Lyde claimed, in a familiar refrain, was really a struggle between Europe (represented by France and Britain) and Asia (represented by Germany). The involvement of Russia on the 'European' side was obviously something of a problem in this respect and was therefore passed over in silence until the Bolshevik Revolution and the subsequent withdrawal of Russia from the war, events which Lyde claimed vindicated his ideas. German central Europe, he claimed, was not really 'European' at all. Here, 'Asiatic patriarchialism' had intruded into an environment of 'European . . . particularism'. The former was 'Oriental despotism' by another name and had spawned a political culture whose 'chief instruments of government were the dungeon and the fortress'. The latter was the basis of democracy, liberalism and personal freedom.[139]

Lyde's rather optimistic conclusion, which sat uncomfortably alongside his hostility to Germany, was that 'scientific frontiers' should be 'anti-defensive' and designed to facilitate contact and co-operation between different peoples and states. Rivers, as 'natural' meeting places, were the ideal international frontiers, affording some protection but ultimately removing the need for such precautions by encouraging trade, exchange and 'a maximum of peaceful tendencies'. Lyde argued that,

> Civilization is progress in the art of living together, and it is long now since the world became an economic unit. . . . We want, therefore, a feature which actually encourages peaceful international intercourse. . . . [T]he river frontier cements economic unity, and ignores dynastic legacies based on the delusion that a river basin must always be an appropriate political unit.

This was no recipe for a 'flabby and maudlin cosmopolitanism', he insisted, but a route to 'the true internationalism of to-morrow' based on 'a stable condition of friendly relations between patriotic and coherent democracies'.[140]

Holdich's view of Germany was no less hostile than Lyde's but this led him to a fundamentally different conclusion. Lyde was absurdly naive, claimed Holdich, and the idea of a political border 'designed so as to give special opportunity for the interchange of social courtesies and the promotion of brotherly good fellowship between contiguous peoples' was laughable, the product of an 'excessively hopeful view of the regeneration of humanity'.[141] Holdich saw no evidence 'of brotherly love . . . [or] yearning loving-kindness amongst peoples who exist . . . as rivals in the great world-field of commercial development and wealth-hunting'. All talk of inter-

national co-operation and a League of Nations was so much hot air: the
only realistic option was a new Europe built on defensible boundaries,
preferably arranged in relation to mountain or upland ranges:

> So long as man is a fighting animal he must be prevented from phys-
> ical interference with his neighbour by physical means ... We must
> reluctantly admit that the best way to preserve peace amongst the
> nations is to part them by as strong and as definite a physical fence as
> we can find.[142]

Where nature was unable to provide a physical barrier, humankind would
have to improvise its own equivalent:

> it seems that the European boundary of the future will be something
> more than the artificial impress of a line ... having no further signifi-
> cance than a hedge. It may well become an actual physical military
> barrier bristling with obstruction and points of steel, so complete and
> effective in its military appointments as to approach ... the ideal of
> absolute security.[143]

Holdich shrugged off the argument that communication between states
with frontiers determined by physical obstacles would be more difficult.
Economic considerations were of secondary importance, he claimed. The
suggestion that new states created on this basis would be no less ethnically
diverse than the existing empires was, Holdich accepted, a more serious
charge. His solution was brutally simple: massive population movements
across newly established frontiers to create as much racial and linguistic
unity as possible.[144] 'Socialist theories of equality unhappily clash with the
hard fact' of Europe's geopolitical antipathies, he observed.[145] Multiracial
states were inherently unstable as different peoples were mutually exclusive:
'an admixture ... may be effected mechanically, but real chemical fusion
never takes place'.[146]

The Holdich–Lyde debate was misleadingly represented as a clash
between a right-wing soldier and 'socialist' academic (Lyde claimed to be a
supporter of the British Labour Party). In Europe, most commentators
sided, somewhat reluctantly, with Holdich.[147] In the USA, however, Lyde's
views were treated more sympathetically, although his eccentric musings
tended to undermine his credibility. The US geographer Albert Perry
Brigham sought to adjudicate between the two perspectives, suggesting that
both kinds of 'scientific frontier' should find their place in a rebuilt Europe.
What was needed, claimed Brigham, was a world survey of boundaries and
frontiers, to begin in Europe. Here, he believed 'twenty-five human groups
... show such unity of purpose and ideal, such community of interest, of
history, and of hopes, and each in such reasonable numbers, that they have
embarked or deserve to embark on a career of nationality'.[148]

A regional utopia

There were, of course, radical European perspectives which rejected all of the above proposals. Some of the most intriguing were developed by a small group of British and French writers and academics who looked forward to what we would now call 'a Europe of the regions'. The regional approach was by no means coherent or consistent but it was always predicated on dissatisfaction with the structure of the existing European nation-states, on both sides of the wartime divide.

Hostility to the European nation-state was widespread amongst left-wing radicals, including most members of the pacifist movement. British 'progressives' such as H.N. Brailsford, J.A. Hobson and E.D. Morel and peace campaigners in the Union of Democratic Control, for example, wrote passionately on the need for international or even global governance, based initially on a united Europe.[149] A defining statement was Hobson's *Towards International Government*, published in 1916.[150] The 'regionalists', though strongly internationalist, pitched their critique at the subnational level. An influential figure here was Patrick Geddes, the remarkable Scottish polymath who wrote prolifically on everything from biology to town planning. Geddes was heavily influenced by the French social critic Frédéric Le Play and the anarchist geographer, Elisée Reclus (whose globe at the 1900 *Exposition* we have already encountered). Geddes travelled extensively, particularly in India, and taught botany (and anything else that came to mind) at Dundee College, then part of St Andrew's University.[151]

Geddes's geopolitical vision stood in marked contrast to the prevailing orthodoxy. Whereas most commentators argued that the existing European nation-states were too small for the coming era, Geddes believed they were already far too large. This had created a kind of spiritual malaise, a pervasive sense of cultural alienation amongst the great mass of the European people. Large states could only be sustained by force, Geddes argued, or by the inculcation of an aggressive, militant nationalism, a 'false consciousness' which had swept aside earlier, more harmonious and collective forms of place-bound identity. A genuinely peaceful geopolitics implied smaller and more meaningful territorial units operating at what was then the subnational, regional scale. These should not be independent microstates, however, but closely interconnected political units. In contrast to most prophets of national self-determination, Geddes rejected race as a legitimate geopolitical category and relied instead on socio-economic typologies, derived mainly from Le Play, which emphasised the formative significance of work and the labour process in creating different kinds of social organisation.

Geddes's wartime ideas were developed in a remarkable series of books, co-edited with his long-time collaborator Victor Branford, entitled *The Making of the Future*. In these volumes, Geddes denounced the

THE WAY OF RECONSTRUCTION (89).

ARMAGEDDON. (A.D. 1914-16).

Figure 2.17 'Armageddon (A.D. 1914–16)' and 'The Way of Reconstruction'. Source: Geddes (1918, 10 and 19).

'Mechanical–Imperial–Financial Age' which had prevailed before 1914 and had propelled Europe towards war. This had not been an era of peace, he insisted, but of 'wardom in thin disguise', an age of spirit-crushing preparation for war. 'The great deficiency of the Mechanical Age', he wrote with his co-author Gilbert Slater, 'is its sacrifice of Life to Things'.[152] The war was 'a gigantic Dance of Death, for which modern business with its associated politics and diplomacy were the long-drawn out rehearsal'.[153] Though terrible in every respect, the war nevertheless represented an opportunity to change the world for the better (Figure 2.1). It marked

> the closing of one era and the opening of another. On the one hand it is the poisonous fruit of an age of pitiless competition and machiavellian diplomacy. But on the other, it expresses a spiritual protest and rebound against the mammoth of materialism. In all its nobler aspects, its heroisms and self-sacrifices, does not war hold promise of renewing Life?[154]

'The coming polity', as Geddes called it, needed a new geopolitical imagination, a communal spirituality derived from small-scale arts and crafts industries and regional forms of production, consumption and governance. 'We are disillusioned with the great nations of imperial aspiration', he claimed with Slater. 'We see the day of the small peoples returning'.[155]

Geddes's vision of a regional Europe was never fully developed though he regularly implied that new territorialities need bear no relation to the continent's existing political geography. In one essay, written for the Cities Committee of the Sociological Society in 1918 on the rival impulses generating war and peace, Geddes included a sketch map of the ferry ports of southern England and northern France displayed in such an orientation as to underline their functional economic similarity rather than their geopolitical difference. Bringing together these ports and their hinterlands within a new regional structure would make more sense, Geddes implied, than separating them in different states (Figure 2.18).[156]

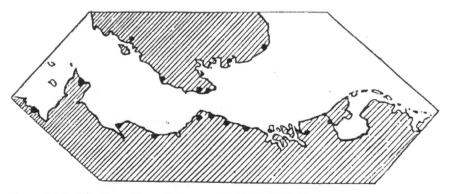

Figure 2.18 'The Ferry Towns'. Source: Geddes (1918, 2).

Geddes's utopian regionalism influenced dozens of more prosaic writers. Newbigin's views, discussed above, owed much to Geddes, as did those of H.J. Fleure, doyen of the Aberystwyth school of geography and anthropology. Fleure's prolific wartime writings on Europe were an eloquent critique of the old imperial nation-states and a vigorous defence of regionalism. Fleure devised a typology of European regions, identifying regions of 'hunger', 'debilitation', 'increment', 'effort', 'industry', 'lasting difficulty' and so on. Each could be developed in its own way by regional government and institutions which could foster a truly humane civic pride, free from the chauvinism and prejudice of the larger nationalisms. Regional government was, Fleure believed, the only alternative to the predatory imperialism which had become the main characteristic of European states before 1914. Like Geddes, Fleure was involved with the Le Play Society and the Regional Survey Association which championed the cause of subnational regionalism.[157] Fleure's approach was endorsed by other academics, notably A.J. Herbertson, Mackinder's successor as Director of the Oxford School of Geography. The 'natural region', Herbertson insisted, was the *genius loci* of all human organisation, the logical building block of a modern and democratic geopolitics.[158]

Converting the regional ideal into practical proposals before and during the war was a difficult task. Arguments for a serious change in the nature of government in the United Kingdom were unlikely to succeed, despite several important advocates. In 1901, Mackinder had argued for northern and southern regional councils in England to reduce its geopolitical dominance over the other British 'nations'.[159] Six years later H.G. Wells proposed a 'new heptarchy' of English regional councils to bring government closer to the people.[160] The onset of war and the nationalist rising in Dublin in 1916 added a fresh urgency to these discussions. The call for 'Home Rule' in Ireland led a few commentators in other parts of Britain to argue for 'Home Rule All Round', a general devolution of political power away from the Westminster-based national Parliament and towards elected regional councils. A federal UK state might then emerge to take its place in a reborn, post-war Europe.

One of the most remarkable British schemes was developed in 1917 by the geographer C.B. Fawcett.[161] Fawcett argued that Britain's national government had become chronically 'overburdened'. The time had come for radical constitutional and administrative reform based on large-scale regional devolution of economic and political power. Ireland, Scotland and Wales should be granted devolved governments, possibly based on more than one regional council. The more serious problem arose in England, the largest of the British 'nations' and the most urgently in need of regional devolution. What would a regionally devolved England look like? Fawcett's initial answer was extremely radical. While recognising the enduring importance of the old English counties in the popular imagination, he advocated a complete break with the past. The war, he argued, had produced a general willingness to rethink conventional outlooks, both at home and abroad. English government should be conducted through 11 provincial councils whose jurisdictions should cut across the old county boundaries. These old divisions would, in time, lose their significance. Fawcett's provinces were designed to reflect England's existing economic and social geography and facilitate balanced post-war economic development. Centres of population and larger drainage basins should, wherever possible, be located within a single province to ensure a minimum population of one million and maximum transport and infrastructural cohesion. His 1917 provinces contained populations ranging from one million in Devon and Cornwall to 6.1 million in Lancashire. Like Geddes, Fawcett argued that regional capitals must be centres of learning and culture as well as economic and administrative foci. Great store was placed on the role of universities as regional centres of innovation in science, technology and the arts. The most radical dimension of Fawcett's 1917 scheme was his suggestion that London be divided between two provinces, along the Thames. The northern part of the city would be the administrative capital of the new London province, which would also incorporate Middlesex, Hertfordshire and much of Essex; south London would become part of the South East England province, whose capital would be Brighton. The objective was to diminish the power of London within the regional tier of government, though, of course, the city would remain the national capital. Fawcett unfortunately toned down this most intriguing aspect of his original proposal in the book which he subsequently devoted to this theme, published in 1919 in the Geddes–Branford series on *The Making of the Future*. Here he argued for 12 rather than 11 provinces and treated London as a single entity, the capital of its own province (Figure 2.19).[162]

Similar ideas were discussed in France where Jean Charles-Brun and an eclectic group of provincial writers, poets and politicians launched the *Fédération Régionaliste Française* (FRF) in 1900 to campaign for regional devolution and to counteract the growing dominance of Paris over the French provinces. The dominant tone of the organisation's monthly journal, *Action Régionaliste*, was left-liberal, but the movement encompassed an extremely wide spectrum of political views. Articles by anarchists such as Élisée Reclus advocating small-scale community life jostled uncomfortably alongside diatribes from nascent fascists such as Charles Maurras and Maurice Barrès who saw France's rural integrity and racial purity challenged by the corrupting influence of cosmopolitan Paris.[163] In 1910, the geographer Paul Vidal de la Blache, whose work on the wartime *Comité d'Études* is discussed above, proposed a new administrative geography for France based on 17 powerful provincial councils which he argued should replace the *départements*, as they were too small to have any serious economic or administrative role (Figure 2.20).[164] Vidal de la Blache's administrative system had more in common with that of the *Ancien Régime* than with that of the nineteenth century. It was, in essence, a proposal to dismantle the administrative reforms introduced after the French Revolution which were designed explicitly to centralise power in Paris (see Chapter 1). The scheme was warmly commended by Charles-Brun and the FRF.[165] In the midst of the war, Jean Hennessy, member of the French Chamber of Deputies, drafted a formal proposal on 29 February 1916 recommending that the nation's administrative geography be altered to conform to Vidal de la Blache's proposal. This was the first serious attempt to reform one of the most centralised European states.[166]

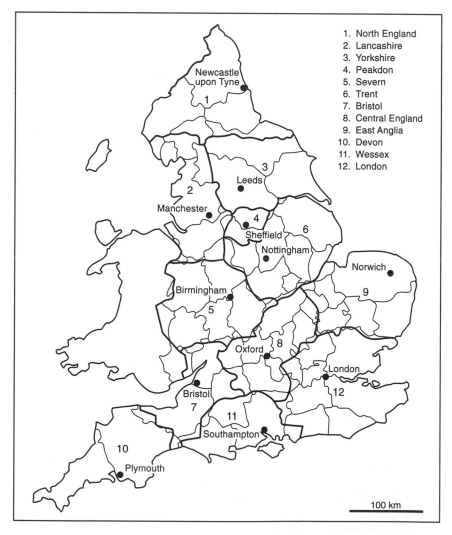

Figure 2.19 C.B. Fawcett's English provinces, 1919. Source: after Fawcett (1919, 248).

The devolution debate in Britain and France sparked off a general debate in both countries about the possibility of an entirely reshaped Europe, in which a confederation of regional governments, each based on a major city, would shape the continent's economic and political agenda in the post-war era. The idea of a devolved regional structure for a post-war Germany was widely debated, though the punitive desire to limit the power of Prussia within a post-war Germany generally outweighed a principled commitment to the inherent benefits of federal government (Figure 2.21).[167]

Figure 2.20 Paul Vidal de la Blache's French regions, 1910. Source: after Vidal de la Blache (1910, opposite 849). Note: the proposed provinces, each with their own capital city, are indicated by the thick lines; the thin lines are the departmental boundaries.

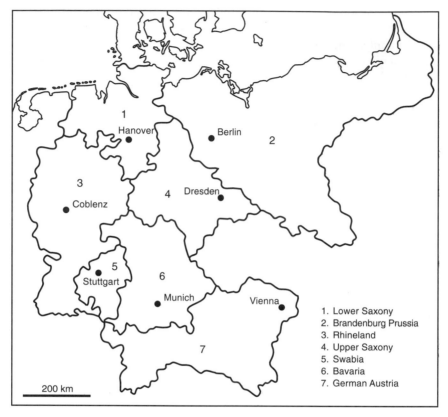

Figure 2.21 Proposed German provinces, 1919. Source: after Anon. (1919, 117).

Summary

The foregoing discussion has reviewed competing ideas about Europe's political geography in the years before and during World War 1. These ideas were connected to wider geopolitical theories and ambitions; all reflected and sustained the *fin-de-siècle* ambivalence about Europe's potential in the twentieth century. Some commentators were upbeat and optimistic, despite the horrors of the Great War and widespread evidence of European decline; others (and they were probably the majority) adopted a more sombre and pessimistic tone. These were by no means the only dimensions to the European geopolitical debate in this period but they give a flavour of the dominant themes. As we shall see in the next chapter, many of these themes remained at the forefront of European geopolitical debate during the 1920s, 1930s and 1940s.

Notes to Chapter 2

1 Cronin (1989, 22–8).
2 Dunbar (1974); D. Gregory (1994, 38).
3 Greenhalgh (1988, 112).
4 This phrase, like the title of this chapter, is stolen from Kearns (1993, 6). See also Laqueur (1996) and Teich and Porter (1990).
5 Nye (1984).
6 McLaren (1983); Offen (1984); Ogden and Huss (1982); J.-J. Spengler (1979).
7 Weber (1986).
8 Quoted in Bayly (1989, 172–5).
9 Schorske (1979).
10 Figes (1996).
11 Djikink (1996, 41).
12 MacKenzie (1984).
13 Even the USA, a nation forged in an anticolonial struggle for independence and blessed with more than enough space of its own, joined the clamour for overseas imperial territory. Following victory in the Spanish–American war of 1898, the USA seized its 'imperial spoils' of Puerto Rico, Cuba and the Philippines.
14 Andrew and Kanya-Forstner (1981, 14–17).
15 F. Fischer (1967, 102–4).
16 Kennedy (1988, 249–354); see also Herrmann (1996); Massie (1991).
17 Winter (1988, 27–41).
18 Adas (1989, 345–401); Kern (1983, 259–86); Pick (1993); A.J.P. Taylor (1969).
19 Heffernan (1996); Robic (1996).
20 Blouet (1987); Kearns (1985); W.H. Parker (1982).
21 Kearns (1997).
22 Mackinder (1887).
23 Mackinder (1911); Ò Tuathail (1996, 75–110); J. Ryan (1994).
24 Mackinder (1904).
25 Mackinder (1904, 421).
26 Mackinder (1904, 421).
27 Mackinder (1904, 422).
28 This challenged the orthodox view, restated in 1890 by the US military historian A.T. Mahan in his *The Influence of Seapower upon History*, that control of the seas was the key to strategic and economic power; see Ò Tuathail (1996, 38–43).
29 Mackinder (1901).
30 Hauner (1992, 142–5). The *fin-de-siècle* enthusiasm for ambitious railway schemes, such as the German Berlin-to-Baghdad project, the British scheme for a Cape-to-Cairo railway and the French project for a north–south Trans-Saharan route, seemed to confirm Mackinder's view.
31 Mackinder (1904, 434).
32 Mackinder (1905).
33 See also Mackinder (1917).
34 On Mackinder's anti-Bolshevism, see Blouet (1976).
35 Mackinder (1919, 161).
36 Mackinder (1919, 178).
37 Mackinder (1919, 194).
38 Mackinder (1919, 155–60).
39 Mackinder (1919, 209).
40 Mackinder (1919, 2–3).
41 Chisholm (1919, 250); Dryer (1920).
42 Mackinder (1904, 441).

43 Teggart (1919, 240–1).
44 Scheler (1915).
45 Brechtefeld (1996); Droz (1960); Le Rider (1994); Meyer (1946; 1955); Schultz (1989); Stirk (1994a); Szücs (1988).
46 Okey (1992, 114).
47 Stirk (1994a, 12; 1996, 10).
48 Stirk (1996, 12).
49 Kann (1964, 157–67); Mommsen (1963); Stirk (1994a, 9; 1996, 13).
50 Fletcher (1984).
51 Partsch (1904).
52 Partsch (1904, 10–11).
53 Quoted in Stirk (1996, 20).
54 Penck (1915).
55 Naumann (1916); see also Hassinger (1917).
56 Bergson (1915).
57 Chisholm (1917a; 1917b); Lyde (1916–17).
58 Verosta (1977).
59 Andrew (1985, 115–35); Pick (1993, 115–35); Richards (1993, 111–52).
60 Sandner (1989).
61 Raffestin *et al.* (1995, 31).
62 Dickenson (1943).
63 W.D. Smith (1980); see also Hunter (1983); Wanklyn (1961).
64 Holdar (1992); see also Ò Tuathail (1996, 43–5); Raffestin *et al.* (1995, 77–102).
65 Ratzel (1897, 2); Bassin (1987a, 480).
66 Herb (1997, 51).
67 Kjellén (1924, 86 and 117); Raffestin *et al.* (1995, 85–6); see also Breuilly (1992).
68 Raffestin *et al.* (1995, 29–75).
69 Ratzel (1900).
70 Ratzel (1901).
71 Neumann (1996) is an excellent account of Russian visions of Europe over the last two centuries.
72 Hosking (1997, 367–97).
73 Chadwick (1945, 117); Kohn (1960).
74 Quoted in Hosking (1997, 368–9).
75 Hauner (1992, 23–4).
76 Bassin (1991).
77 Bassin (1994).
78 Bassin (1993).
79 Quoted in Hauner (1992, 3).
80 Hauner (1992, 155).
81 Hauner (1992, 46).
82 Bishop (1989); French (1995).
83 Curzon (1899); quoted in Hauner (1996, 46).
84 E. Huntington (1907).
85 Bassin (1993).
86 Curzon (1907).
87 Kearns (1984).
88 Banks (1996); Banton (1987); Stepan (1982); Stocking (1987).
89 Mosse (1978).
90 Livingstone (1992, 216–59).
91 Biddis (1970). On the fear of racial degeneration, see Chamberlin and Gilman (1985) and Pick (1989).

92 Peet (1985).
93 Rich (1984).
94 Paul (1984).
95 Mosse (1978); Poliakov (1974); see also Chamberlain (1915); Clarke (1916).
96 Ripley (1899).
97 Slezkine (1994).
98 Crook (1994); Eksteins (1989, 90–140); Hynes (1990, 3–24); Pick (1993, 136–64); Stromberg (1982).
99 Stepan (1987); see also H. Campbell (1918); Mitchell (1915).
100 Holdich (1916a, 15 and 245).
101 Quoted in Pick (1993, 83); see also Darwin (1926); Jordan (1915).
102 Renan (1990); see also A.D. Smith (1986; 1991).
103 Barker (1915; 1918).
104 Barker (1914); Munro (1919); Pick (1993, 88–96).
105 A good example is Nalkowski (1917).
106 See, as one example, Reclus (1914).
107 Hettner (1915a; 1916; 1917a; 1917b); Penck (1916).
108 Hettner (1915b, 28); also quoted in F. Fischer (1967, 159–60).
109 Hettner (1917a).
110 Grant (1916); see E. Huntington (1915; 1919; 1924).
111 D.W. Johnson (1915; 1918; 1921).
112 D.W. Johnson (1917b; 1917c).
113 D.W. Johnson (1917b, iv).
114 Gelfand (1963); Herb (1997, 13–33).
115 Seymour (1951, 2).
116 Seymour (1951, 3). Bowman's work was supported by several leading US geographers, including Douglas W. Johnson and Mark Jefferson, who were recruited as the expert on boundaries and the chief cartographer, respectively, as well as Nevin Fenneman and Ellen Churchill Semple.
117 Dominian (1917, vii).
118 Dominian (1917, 4). The persistence of these ideas can be detected in Cornish (1936) and Chadwick (1945).
119 Henig (1995, 75–6).
120 Masaryk (1918).
121 H. Seton-Watson (1917).
122 A. Evans (1916, 262); see also H.C. Woods (1919).
123 Cvijic (1916; 1918).
124 Newbigin (1915; 1917).
125 Newbigin (1918).
126 Newbigin (1917, 320). The idea of Newbigin as an early prophet of multiculturalism cannot be taken too far as she also betrayed more than a hint of antisemitism which she mixed, as was so often the case, with fear of Communism. She believed Jews were too 'internationalist' ever to enter fully into the community of a nation. The Jew, she claimed, in a review of Nalkowski's 1917 plea for a new Poland, was 'a dangerous citizen in a nascent State'. Jewish internationalism, she believed, was directly connected to 'that body of social doctrine to which the vague name of Bolshevism is given ... the product of the Ghetto and the sweating shop. ... [Bolshevism] is largely generated among the poorer Jews on the western margins of Russia'; see Newbigin (1919, 91).
127 Heffernan (1996).
128 F. Fischer (1967); Koch (1984); Stevenson (1982).
129 Vidal de la Blache (1903); see also Robic (1994).
130 Digeon (1959).

131 The idea of rural settlement patterns as an indicator of race (and hence nationality) was a common feature of many early historical geographies of Europe and was retained as a heuristic strategy in mid-twentieth-century accounts; see Houston (1953) and the discussion in Murphy (1990).
132 *Comité d'Études* (1918–19).
133 Launay (1917, 251).
134 Vidal de la Blache's *La France de l'Est*, published in 1917 (the year of his death) and often regarded as his greatest achievement, was largely the product of research carried out for the *Comité d'Études*; see Gallois (1918).
135 Curzon (1907, 2 and 53–4); see also Kristof (1959); Prescott (1978, 1–32).
136 Lyde (1913).
137 Holdich (1909).
138 H.J. Johnson (1919).
139 Lyde (1915, 33–5; 1916). Lyde's hysterical anti-German beliefs reached surreal levels in his *Peninsular Europe* of 1931 in which he refers to the tell-tale 'Prussian squint ... [or] slightly oblique eye', undeniable evidence of 'Tartarisation' which the *Junker* class sought to disguise by growing the familiar flamboyant moustaches; see Lyde (1931, 288–9).
140 Lyde (1916, 548–55); see also Lyde (1914; 1915; 1919).
141 Holdich (1916b, 421–2).
142 Holdich (1916c, 498).
143 Holdich (1916c, 507).
144 Holdich (1916a; 1918a).
145 Holdich (1918b, 3).
146 Holdich (1916c, 499).
147 Fawcett (1918).
148 Brigham (1919, 218); see also D.W. Johnson (1917a).
149 Ceadel (1987); Swartz (1971).
150 Hobson (1916); see also Brailsford (1914); Hobhouse (1916); Morel (1916; 1917).
151 Mellor (1990).
152 Geddes and Slater (1917, 188).
153 Geddes and Branford (1916–17, 100).
154 Geddes and Branford (1916–17, 100).
155 Geddes and Slater (1917, 200); see also Branford and Geddes (1919a).
156 Geddes (1918b). This image was one of 94 gloriously eclectic photographs and paintings, mostly produced by Philip Maigret, Geddes's first biographer, which were subsequently made into lantern slides and offered for hire, along with the text of the paper, to clubs and societies around the country who wished to debate the implications of the Geddesian world-view; see also Geddes (1915); Branford and Geddes (1919a; 1919b).
157 Fleure (1915; 1916; 1916–17; 1917; 1918); see also Gruffudd (1994); Matless (1992).
158 Herbertson (1905; 1913; 1913–14); see also Unstead (1916).
159 Mackinder (1901, 122); see also Mackinder (1919, 253–5).
160 Wells (1908, 274).
161 Fawcett (1917).
162 Fawcett (1919); see also Harvie (1991).
163 Flory (1966).
164 Vidal de la Blache (1910).
165 Charles-Brun (1911).
166 Hayward (1973); Mény (1974).
167 Anon. (1919); Peake (1919); see also E.W. Gilbert (1952); Harvie (1994).

|3|

Land and power: the geopolitics of peace and war

Introduction

This chapter considers the evolution of the European idea from 1918 to 1945 with reference to the three rival ideologies which dominated European politics in this period: liberal democracy, communism and fascism. As we shall see, these were by no means internally consistent world-views. They represented distinct but complex ideological currents, each with their own substreams, eddies and counter-flows. Together they provided the context within which a new European geopolitics arose. The collision between these three ideologies reflected a struggle over the meaning of Europe which culminated with World War 2.

After the deluge . . . business as usual

The politicians, diplomats and academics who assembled in Paris for the peace conferences did so in the knowledge that the preceding four and a half years of war had claimed an unprecedented number of European lives. Precisely how many was a matter of dispute, then as now, even in respect of the military casualties.[1] We now know that around eight million European combatants were killed, twice the number of deaths in all previous wars between 1789 and 1914.[2] This represents an average casualty rate of over 5000 soldiers per day throughout the war. In absolute terms, Germany suffered most, losing 1.8 million soldiers. Next came Russia with 1.7 million, France with almost 1.4 million, Austria–Hungary with 1.3 million and Britain with nearly 750 000. Proportionately, the worst affected country was France, where one in four male children born between 1891 and 1895 had died in battle by November 1918.[3] Entire armies were decimated. Serbia lost almost half of its military force, Turkey and Rumania perhaps a third and Bulgaria almost a quarter.[4] The vast majority were young men in their 20s; some 63 per cent of Germany's war dead had been in this age-group in August 1914.[5] Most were conscripts rather than professional soldiers; between 12 and 17 per cent of the adult male population of Europe were conscripted.[6]

Civilian death rates were also exorbitant. German civilians deaths from disease and malnutrition equalled British military casualties. Serbia's civilian deaths outnumbered even the huge losses from the country's army. At least a million Armenians were murdered in Turkey in the twentieth century's first genocide. To make matters even worse, a lethal influenza pandemic swept around the world in 1918 claiming more lives than the war.[7] Globally, perhaps 50 million civilians died from disease and starvation.[8] In parts of Europe, 20 per cent of the labour force were wiped out.[9] To all this must be added the tragically broken and brutalised survivors of the war, at least 20 million souls, whose minds and bodies had been permanently shattered. Many of these victims joined the post-war armies of shuffling beggars and drifters, eking out a meagre existence on the mean streets of Berlin, Paris, London and Vienna.[10]

The war dead cast a giant shadow over inter-war Europe. Photographs of uniformed young men nestled on cabinets and sideboards in the grandest mansions and the humblest cottages while entirely new official forms of collective remembrance developed at civic monuments and memorials across the entire continent to recall their sacrifice.[11] The ghostly presence of the war dead convinced many that 1914–18 marked a fundamental historical discontinuity in which the values and attitudes of the nineteenth century were utterly destroyed.[12] The hauntingly beautiful words of Dick Diver, the central character in F. Scott Fitzgerald's *Tender is the Night* (1934), encapsulate the viewpoint of the survivors:

> This western-front business couldn't be done again. ... The young men think they could do it but they couldn't. They could fight the first Marne again but not this. This took religion and years of plenty and tremendous sureties and the exact relation that existed between the classes. ... You had to have a whole-souled sentimental equipment going back further than you could remember. You had to remember Christmas, and postcards of the Crown Prince and his fiancée, and little cafés in Valence and beer gardens in Unter den Linden and weddings at the mairie, and going to the Derby, and your grandfather's whiskers ... and country deacons bowling and marraines in Marseilles and girls seduced in the back lanes of Württemberg and Westphalia. Why, this was a love battle – there was a century of middle-class love spent here. ... All my beautiful lovely safe world blew itself up here with a great gust of high explosive love.[13]

The ghosts of Europe's missing generation bore down on the shoulders of political leaders poring over their maps and documents in Paris in 1919. But their strategies and negotiations suggest that the war had not radically shaken the conventional forms of European geopolitical thinking.[14] Rather than enthusiastically embrace the mood of change, most political leaders sought to limit and control the process of geopolitical transformation which was already underway, symbolised by the collapse of the great autocratic monarchies of central and eastern Europe, in Russia following the Revolution in 1917 and in Germany as an Allied precondition for the November 1918 armistice.

Figures 3.1 and 3.2 show the main changes to Europe's political geography in respect of Germany and Austria–Hungary, the former following the Treaty of Versailles (28 June 1919), the latter following the Treaties of St Germain, Neuilly and Trianon (signed with the new governments of Austria, Bulgaria and Hungary on 10 September 1919, 27 November 1919 and 4 June 1920, respectively).[15] The impressive scale of these changes cannot disguise the fundamentally conventional nature of the geopolitical motives which informed this restructuring. This is not to say that no one was interested in devising a new European order which might deliver long-term peace; it was simply that few of those capable of influencing events were able to think beyond the conventional forms of national and imperial geopolitics which had brought about the war in the first place. The two main concerns of Allied strategists – the desire to punish and control Germany and the perceived need to contain or destroy the threat of Bolshevism in its unsteady Russian power base – reflected traditional geopolitical thinking. Both these objectives presupposed some form of territorial restructuring to shift the balance of power in favour of the victors. This was the conventional result of an episode of war, one which would have been entirely familiar to those gathered at the Congress of Vienna in 1815. The new European map reflected a conservative desire to reorder the political landscape to reward the victors, to punish the losers and to head off the threat of revolution.[16]

Advocates of national self-determination on racial or linguistic grounds, including several US delegates and the representatives of minority peoples, insisted that their strategy marked a genuine departure from older forms of imperial geopolitics. In one sense, this was quite true. As we noted in Chapter 2, traditional methods of post-war territorial readjustment by direct land exchanges between 'winners' and 'losers' simply reinforced an imperialist lust for space, a desire to claim 'the spoils of war'. The 'new' idea of granting national rights to racial minorities challenged the very idea of the imperial nation-state and was often inspired by a genuine conviction that small states were more democratic than larger ones.[17]

Unfortunately, self-determination was a far less radical alternative than its proponents argued. It too was based on the assumption that geopolitics must be defined by conventional nation-states which, in turn, were assumed to exist only as discrete plots of territory. In this sense, self-determination can be seen as the culmination of territorial politics and all the more

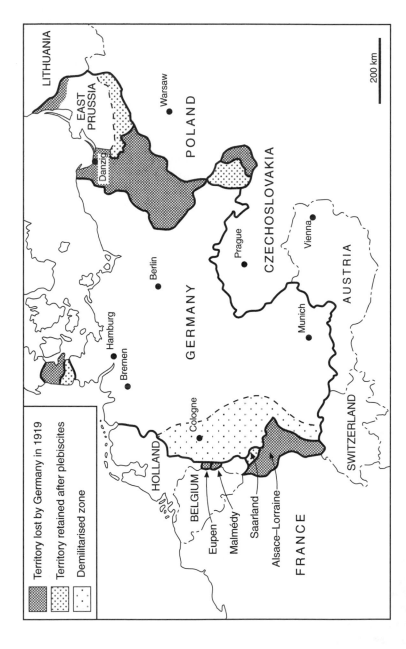

Figure 3.1 Germany after the Treaty of Versailles, January 1920.

Figure 3.2 The end of empire: the break-up of Austria–Hungary.

dangerous for being grounded in the belief that 'race' determines nationality, a particularly corrosive idea which was to reach its perverted heights during the 1930s and 1940s.

Self-determination was, in any case, destined to have limited influence on decisions in Paris. Britain was still in the midst of an ongoing crisis with Irish nationalists whose arguments for independence, violently asserted and crushed during the Dublin rising of Easter 1916, drew explicitly on the idea of self-determination for small nations.[18] The Irish problem raised a dilemma for the British government: an enthusiastic endorsement of self-determination in London would play directly into the hands of those seeking Irish independence.

France was equally ambivalent. The French Prime Minister Georges Clemenceau was the principal advocate of a harshly punitive peace treaty with Germany and did not wish this simple objective to be complicated by side issues relating to the rights of small nations, unless, of course, this helped his main objective. As imperial powers, neither Britain nor France were willing to contemplate self-determination outside Europe and the 'white' settler colonies.

Self-determination was therefore only applied selectively, either to re-inforce the direct assault on the territorial integrity of the Central Powers and Bolshevik Russia or to create 'buffer-states', dependent on 'western' European support. These states were to prevent future German expansion towards the east (Poland) and south (Czechoslovakia and Yugoslavia) or future Russian expansion towards the west (Poland, Finland, Estonia, Latvia, Lithuania). 'Buffer-states' needed to be strong enough to perform this task. In the cases of Czechoslovakia and Yugoslavia, this demanded fed-eral, multi-ethnic arrangements rather than racially cohesive geopolitical units. Yugoslavia's ethnic composition was particularly complex: 39 per cent of the population were Orthodox Serbs, 24 per cent Catholic Croats and the remaining 37 per cent included at least 15 other ethnic groups. These figures are only estimates, however, for the ethnic complexity of the region had baffled most delegates at the peace conferences. Experts on the region, such as the Paris-based Serbian geographer Jovan Cvijic (whom we encountered in Chapter 2), produced spurious ethnic maps exaggerating the number and distribution of Serbs. These were designed to enhance Serbian dominance over the new federal state. Sympathy for the Serbian cause was running high in 1919 in view of the country's terrible suffering. As a result, most commentators happily endorsed the idea of a 'greater Serbia' within a federal Yugoslavia.[19] H.V. Temperley, the official British historian of the peace conference detected 'obvious traces of the hand of . . . Cvijic' in the Yugoslav federal structure, and Peter Taylor has claimed that '[t]here is probably no other example of one man influencing national definition so completely as Cvijic'.[20] Counter-propaganda opposing the new Yugoslav state on the grounds that it compromised the rights to self-determination of the smaller peoples in the area (including the people of Montenegro who had their own independent state in 1914) was markedly less successful (Figure 3.3).[21]

The map of Europe fashioned in Paris in 1919 was the outcome of two logically inconsistent strategies deployed selectively by the victorious Allies to achieve highly conventional geopolitical objectives. Germany and Russia were penalised territorially and Austria–Hungary and the Ottoman Empire completely dismembered in the name both of a traditional, punitive geopol-itics and in order to allow national self-determination for minority peoples. The latter justification was, to say the least, self-serving and duplicitous, for this 'right' was permitted only in relation to territory formerly controlled by the Central Powers. This was a geopolitical fix, an attempt to satisfy as many Allied constituencies and nations as possible but which ended up sat-isfying virtually no one and generating immediate instability.

Figure 3.3 'Off the Map': the fate of Montenegro. Source: Devine (1921, frontispiece).

It had been agreed at the beginning of the peace conference in February 1919 to establish a new League of Nations in Geneva as a forum for international debate, a mechanism through which old and new states could negotiate openly and in good faith. Unfortunately, this was unable to prevent the immediate resumption of hostilities in different parts of Europe. This was scarcely surprising for it had barely begun operating when violence flared once again. Yet its potential influence had been fatally undermined from the outset. The rejection of German and Russian involvement hardly reflected the spirit of internationalism which the League had been established to foster. This ideal was further undermined by the allocation of League of Nations 'mandates' to Britain, France and the British Dominions to administer former German and Ottoman colonial territories 'in sacred trust for the good of civilization'. J.A. Hobson dismissed this as a 'veil for the annexation of enemy countries and the division of the spoils', old-style imperialism disguised by 'an increasingly elaborate fig-leaf'.[22]

The fact that the League's chief advocate, the US President Woodrow Wilson, was unable to convince his country's elected leaders of the League's merits was also a serious drawback. Wilson had trumpeted the League in the 14th and final point of his famous declaration on 8 January 1918 (see Chapter 2). By the time he set sail for Paris aboard the *USS George Washington* with his advisers and three army truck loads of House Inquiry reports, statistics and maps, the League was his main objective.[23] Wilson, the first US President to travel outside the USA during his term of office, gathered 12 of his scholarly advisers, including Isaiah Bowman, onto a windswept deck on 10 December. 'Tell me what's right', he demanded, 'and I'll fight for it. Give me a guaranteed position'.[24] And fight he did once he had arrived, particularly for the League, much to the irritation of sceptics like Clemenceau.

Even as agreement was reached, Wilson knew his policy had been undermined by domestic political changes in the USA. The Republican opposition, isolationist and hostile to Wilson's idealistic belief in America's role as an international arbitrator, had gained the majority in both Houses of Congress following elections on the eve of the armistice. A 'Wilsonian' peace treaty, based on a League of Nations which would bind the USA to Europe into the future, would never be ratified in Washington, particularly as the Foreign Relations Committee was chaired by Wilson's arch enemy, Henry Cabot Lodge. The US entry into the war in 1917 had been the clearest rejection yet of the self-imposed geopolitical constraints of the so-called 'Monroe Doctrine' (1823) which had limited US interests to the Americas, north and south, through most of the nineteenth century. This was to be a short-lived experiment: the League of Nations, so ardently championed by the wartime US administration was firmly rejected by its isolationist post-war successor along with the rest of the Treaty of Versailles, which failed to win the necessary two-thirds majority of the US Senate.[25] The League therefore fell under the influence of the old imperial powers, Britain and France, whose attitude towards it had always been lukewarm.

In most parts of Europe, old grievances were settled in the old ways. In the southeast a bitter war erupted in 1920 between Greece and reinvigorated Turkish troops, the latter struggling under the secular leadership of Kemel Ataturk to reassert a nationhood denied them completely by the Treaty of Sèvres (signed on 10 August 1920). The war, which led to massacres, the evacuation of 1.3 million Greeks from around Smyrna (modern Izmir) and enduring mutual hatred, ended when the Treaty of Lausanne established the modern state of Turkey in early 1923. Italy, under the new fascist leader Benito Mussolini, exploited the chaos to invade Corfu in the same year. Italian empire-builders felt themselves badly treated by their wartime Allies who had promised so much to tempt them away from their pre-war alliances with Germany and Austria–Hungary only to offer unimpressive territorial rewards in the shape of the south Tyrol and the Triestino. Corfu was the merest pinprick compared with the huge Italian empire which had been imagined in the Adriatic and Asia Minor, but its seizure indicates the persistence of Italy's geopolitical ambition.

In eastern Europe, the Bolshevik victory in the Civil War culminated with the reconquest of those areas which had seized the opportunity of independence before the end of the war: Armenia (independent 1918–21), Azerbaijan (1918–20), Georgia (1918–21), Ukraine (1917–20) and White Russia (1919–21), costly victories which added to the staggering death toll produced by warfare, terror, famine and disease in Russia after 1917. Most estimates suggest that around 10 million Russians died, substantially more than were killed in all countries (including pre-revolutionary Russia) during World War 1.[76] Frontier disputes between Russia and Poland flared immediately, culminating with a Polish invasion of Russia followed by a Soviet invasion of Poland. The British Foreign Secretary, Lord Curzon, proposed an ethnic division between Poland and Russia, the so-called 'Curzon line', at a conference in Spa in July 1920 (a division which conforms roughly to the modern-day border between Russia and Poland). This was rejected by both sides and the frontier finally agreed at the Treaty of Riga three months later gave Poland 130 000 square kilometres to the east of the 'Curzon line' together with substantial Russian and Ukrainian minorities. Poland annexed the city of Vilna (Wilno to the Poles; modern Vilnius) from Lithuania, 'confirming' its authority over the area by a plebiscite two years later. Under the terms of the Treaty of Versailles, Poland had also been granted access to the Baltic by means of a 'corridor' through the former East Prussia towards the port of Danzig (modern Gdansk), a city administered by the League of Nations. The 'Danzig corridor' gave Poland a huge German ethnic minority. By 1921, the new state of Poland, which had been inaugurated amidst anti-imperial claims of self-determination, possessed a population which was only two-thirds Polish, the rest being German, Russian, Ukrainian and Lithuanian.

Even the countries which fared relatively well from the initial agreement in Paris were worried by the outcome. The enlarged Rumania, which had acquired Transylvania from Hungary as well as great swaths of former Russian and Austrian territory as a 'reward' for its loyalty to the Allied cause, now had to deal with large and troublesome Ukrainian, Hungarian and Moldovian minorities. France, which reclaimed Alsace–Lorraine, was as obsessed as ever by the threat from beyond the Rhine.[27] A massive fortified frontier was planned, the 'Maginot line', to protect *la mère patrie*.[28] Money flowed into the reclaimed provinces to enhance their Frenchness. Urban planning sought to efface the Germanic heritage, and Strasbourg became second only to Paris as a centre of proselytising French civilisation, its magnificent university the focus of the famous *Annales* school of history, founded in the 1920s by Marc Bloch and Lucien Febvre.[29]

The French Foreign Ministry sought anxiously to consolidate its 'success' in imposing a demilitarised 'exclusion zone' in the German Rhineland by funding the small Rhenish separatist movement.[30] Some French strategists hoped that the complex 'federal' structure of post-war Germany, whose capital was now the innocuous town of Weimar rather than the imposing imperial city of Berlin, might yet disintegrate (Figure 3.4). Clemenceau insisted that Germany must accept full responsibility for the war and pay the reparations bill, in cash and material, over the next 30 years. Failure to meet these harsh demands led to a French occupation of the Ruhr industrial belt in January 1923, a disastrous intervention which set back Franco–German *détente* by several years.[31]

The eventual resolution of this crisis, involving a rescheduling of German repayments under the so-called Dawes Plan, backed by a less isolationist US government, finally laid the foundations for wider European stability under the terms of the Treaty of Locarno, signed in December 1925 and guaranteeing the borders of Germany, France, Belgium, Great Britain and Italy. This was reinforced by a complex web of alliances between the new states as they sought security from each other and their more powerful neighbours, a process inaugurated by the 1921 'Little Entente' between Czechoslovakia, Yugoslavia and Rumania.[32] The enhanced stability of the mid-1920s paved the way for Germany's entry into the League of Nations in 1926.

Despite hopes that a more united Europe might emerge from the carnage, the 1925 map reveals a continent more complex and geopolitically divided than it had been in 1914 (Figure 3.5). The main geopolitical surveys of the new Europe, Isaiah Bowman's *The New World* and Jean Brunhes and Camille Vallaux's *La géographie de l'histoire*, show a continent with an extra 20 000 kilometres of international frontier, the product of duplicity and compromise in Paris and the subsequent episodes of warfare and international intrigue.[33]

These new frontiers could only be traversed with the aid of passports and a plethora of official paperwork. The age of strict border controls and

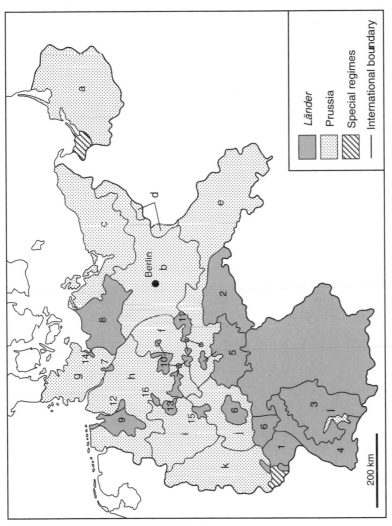

Figure 3.4 Weimar Germany: the federal system. Note: numbers 1–16 refer to the non-Prussian *Länder*: (1) Bavaria; (2) Saxony; (3) Württemberg; (4) Baden; (5) Thuringia; (6) Hesse; (7) Hamburg; (8) Mecklenburg–Schwein; (9) Oldenburg; (10) Brunswick; (11) Anhalt; (12) Bremen; (13) Lippe; (14) Lübeck; (15) Waldeck; (16) Schaumburg–Lippe. Prussia, the largest *Länder*, was made up of several provinces: (a) East Prussia; (b) Brandenburg; (c) Pomerania; (d) Grenzmark–Posen Westpreussen; (e) Silesia; (f) Saxony; .g) Schleswig–Holstein; (h) Hannover; (i) Westphalia; (j) Hessen–Nassau; (k) Rhine; (l) Hohenzollern. Source: Bullock (1991, 151); Freeman (1995, 12).

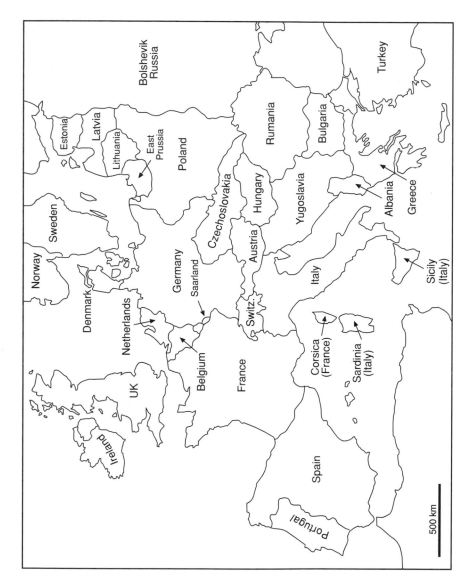

Figure 3.5 Europe after the Treaty of Locarno, December 1925.

immigration checks had begun. This was a direct response to the huge refugee problem created by a political geography which left *irridenta* and ethnic enclaves scattered across the entire continent, a problem which the warfare of the early 1920s exacerbated rather than solved. Between 4 and 5 million people were made homeless or 'stateless' by warfare and border changes.[34] All over Europe, people abandoned villages and towns where they had lived for generations to seek a new life in a reconstituted homeland (Table 3.1).

The new Europe was economically weaker too. The war had cost the equivalent of seven times the world's total national debt since 1700 and the material damage set back real levels of European industrial output by at least a decade. In 1912, Europe had accounted for 43 per cent of world industrial production and 59 per cent of world trade; by 1923, these figures had fallen to 34 per cent and 50 per cent, respectively. The biggest beneficiary of this collapse was the USA – a debtor nation in 1914, the world's largest lender in 1919.[35]

The economic and financial viability of the new European states had been scandalously ignored in Paris. No provision had been made for central banks or national currencies and the infant states faced huge difficulties creating reliable banking structures, money supplies and systems of exchange. Twice the number of pre-war currencies existed (27 compared with 14), making European trade more complex and increasingly prone to tariff adjustments, economic nationalism, currency fluctuations and inflation.[36]

The structure of the transport infrastructure also received little attention in Paris although many states found themselves cut off from traditional markets and resources by new international frontiers. Austria, a shadow of its former extent, was left with a bloated capital city which had previously served a vast empire while some new states had no obvious capital at all, only rival towns occupied by differing ethnic groups. The huge numbers of war wounded, widows and fatherless children also placed a great strain on rudimentary welfare provisions.

Table 3.1   Post-1918 population movements (source: Barraclough, 1984, 265)

From	To	Estimated Number
Turkey, USSR	Greece	1 350 000
Poland, Czechoslovakia, Alsace–Lorraine, former colonies	Germany	600 000
Greece, Bulgaria, USSR, Yugoslavia	Turkey	600 000
Yugoslavia, USSR, Czechoslovakia	Hungary	400 000
USSR	Baltic States	200 000
Greece, Rumania, Yugoslavia	Bulgaria	150 000
USSR	Poland	100 000

The job of repairing war damage was another problem. This had been considered in 1919 only as part of the guilt and reparations clauses inserted in the Treaty of Versailles to punish Germany. No detailed plans were laid down and the response was extremely variable. In the west, the material damage in Belgium and eastern France was enormous but relatively concentrated geographically and was put right with impressive energy.[37] Elsewhere in Europe, particularly in the poorer east, war damage took much longer to repair. In a famous diatribe, the British economist John Maynard Keynes viciously attacked those who had planned the new Europe without reference to such fundamental economic considerations.[38]

Although World War 1 transformed the continent's political geography it did not fundamentally challenge the conviction that Europe was merely a complex system of alliances binding together rival nation-states, each based on discrete territories and defensible frontiers. The position of borders and frontiers had changed but the idea of what a border or frontier represented remained the same. Winston Churchill summed up more accurately than he realised in 1914 the nature of Europe's geopolitical response to the catastrophe through which it was to pass: 'The British people have taken for themselves this motto', he said, 'Business carried on as usual during alterations on the map of Europe'.[39]

After Locarno, optimists hoped the worst was over. The new Europe, though obviously less pivotal in world affairs, could perhaps look forward to a peaceful and prosperous future. But the continent's relative economic decline continued through the mid-1920s. Europe's share of world trade fell by 1 per cent each year, with Britain and Germany leading the decline. Only 45 per cent of global trade passed through Europe in 1928.[40] The pattern of global lending had also been profoundly altered by the war. The USA was now the dominant source of credit, its banks supplying two-thirds of global investment between 1919 and 1929. Europe, Latin America and the European colonies became debtor regions. This system generated a major 'debt crisis' (comparable in some respects with that of the late 1970s and early 1980s) in which widespread defaulting on repayments led to an unprecedented collapse of the US stock market in 1929. The first truly global recession had begun and was to persist for the next five years.[41] The impact of the depression on the European economies was devastating but highly variable. Recovery was also slow and patchy. Overall, the eastern and central parts of the continent fared worst.[42] Europe's increased political fragmentation was worsened by the highly regional impact of the world's most serious economic slump during the early 1930s.

The depression justified the criticisms of Keynes, and his views on strategic economic planning and government intervention to secure maximum employment became hugely influential on both US and European governments seeking a route out of the crisis. But the depression also seemed to vindicate those pessimists who saw Europe as a doomed continent. The anxiety and gloom which had been so prevalent at the *fin-de-siècle* had, if anything, intensified by 1930, fuelled by sombre philosophical works such as Oscar Spengler's *The Decline of the West*, the title of which says it all.[43] The old, pre-war Europe of empires and autocracies had gone but there was little faith that the post-war geopolitical structure was a more viable alternative.[44]

Europe reborn: the pan-Europa movement

Most liberal political leaders were content to muddle through the crises of the early twentieth century, but a vocal minority looked forward to a reborn, united Europe which could retain the liberal and democratic systems of government won at such cost during the nineteenth century.[45] Richard Coudenhove-Kalergi's Pan-European Union (PEU), established in Vienna in 1923, was easily the most successful inter-war organisation devoted to the cause of European unity. The urbane and multilingual Coudenhove-Kalergi (a native of cosmopolitan Bohemia, raised in Vienna by an Austrian diplomat father and a Japanese mother) had initially been an eloquent advocate of the League of Nations but, like many Europeans, he was soon disillusioned.[46] A global organisation was too ambitious, he believed. What Europe needed was a clear strategy to halt and reverse its eclipse and to prevent war in the future. This depended on fostering a new economic and political unity, the only guarantee of prosperity and peace.

Coudenhove-Kalergi's approach was a direct continuation of the pan-regionalism which had flourished in Germany and much of central Europe before the war.[47] He argued that:

> The cause of Europe's decline is political, not biological. Europe is not dying of old age, but because its inhabitants are killing and destroying one another with the instruments of modern science ... The peoples of Europe are not senile – it is only their political system that is senile. As soon as the latter has been radically changed, the complete recovery of the ailing continent can and must ensue.[48]

Drawing on technological, infrastructural and economic arguments laid down over a century earlier by Saint-Simon (see Chapter 1), the PEU emphasised the role of modern communications, especially rail, road and air power, in the creation of new pan-regional systems around the world. This echoed the views of Mackinder, discussed in Chapter 1, but unlike the pessimistic Briton who saw pan-regions as heralding the end of Europe, Coudenhove-Kalergi was convinced Europe would re-emerge as an invigorated and united arena to take its place alongside the other pan-regions, probably focused on the USA, Japan, Russia and the British Empire. A united Europe, based explicitly on a US federal model, would logically dominate Africa to the south, creating thereby a single 'Eurafrican' zone (Figure 3.6).[49]

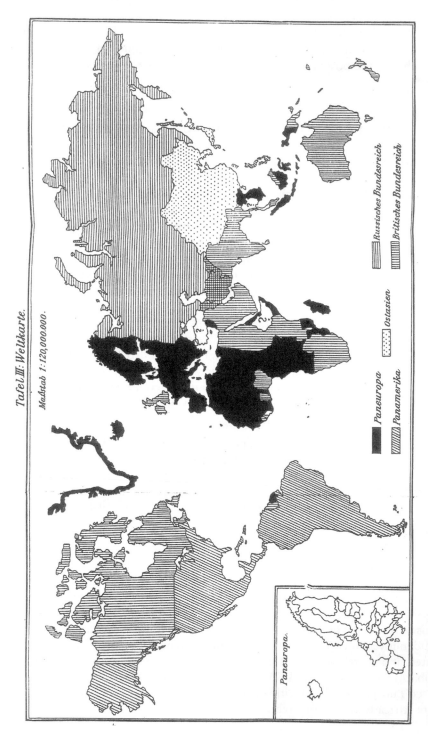

Figure 3.6 A liberal perspective: Richard Coudenhove-Kalergi's *Pan-Europa*, 1923. Source: Coudenhove-Kalergi (1923, 176).

Coudenhove-Kalergi was a convinced multiculturalist. He believed that Europe, rather than the USA, was the original ethnic 'melting pot'. The notion of racial purity, or of racial definitions of nationhood and 'Europeanness', were morally and intellectually redundant, he argued. The European 'family' of nations had emerged in a remarkably short time and had largely served the interests of political élites rather than the mass of the European people. The time of the old nation-states had passed; a new geopolitics of unity geared to the needs of all Europe's inhabitants was required.

For Coudenhove-Kalergi, Europe's limits were flexible and depended not on ethnic, religious or cultural characteristics but on the nature of the political system. The PEU sought a liberal and democratic Europe, safe from anti-democratic ideologies, particularly communism but subsequently fascism.[50] Coudenhove-Kalergi saw no place for Russia in the European system while it remained under the Bolshevik 'yoke': 'History gives Europe the following alternatives', he wrote in 1923, 'either to overcome all national hostilities and consolidate in a federal union, or sooner or later to succumb to a Russian conquest. There is no third possibility'.[51] More surprisingly, perhaps, the PEU rejected the idea that imperial Britain could play a direct role in a united Europe. Britain's place in a new Europe depended on shaking off its imperial traditions and aspirations. Freed from its imperial past, the global experience of the British political élite, and their transatlantic affinities with the USA, would make them the obvious leaders of a united Europe.[52] The PEU was not anti-colonial in principle (as its advocacy of 'Eurafrica' demonstrated), it simply argued for a more rational distribution of colonial responsibilities and advantages between the emerging pan-regions. Britain's chaotically scattered empire simply did not fit with such 'tidy-minded' geopolitics.

Coudenhove-Kalergi's PEU has been variously described as an 'astonishing mixture of large-scale Utopianism, potent political analysis and clear-sighted pragmatism' and 'a blend of fashionable geo-political calculation and an eclectic cultural anthropology'.[53] The organisation won widespread support all over Europe, opening branches in every major city and organising a series of well-attended congresses, beginning in Vienna in 1926. Many academics were converted to its cause. The French geographer Albert Demangeon, an influential member of the *Service géographique française* which had advised the French delegation at the peace conferences, argued that the political system created in 1919 would simply hasten Europe's decline.[54] The only solution was European unity emerging step by step from smaller regional trading blocks, notably in the Balkans, the Danube basin and the Rhine valley. Like Coudenhove-Kalergi, Demangeon believed that Britain would need to dismantle its Empire in order to enter more fully into the European project.[55]

More importantly, the PEU also gained some heavyweight political converts in the shape of Aristide Briand and Edouard Herriot in France (variously Prime Minister and Foreign Minister); Gustave Streseman (Chancellor of Weimar Germany); Edvard Beneš (the senior Czech politician); and even Winston

Churchill in Britain. Churchill was a convinced imperialist, of course, but spoke frequently during the 1920s and 1930s of the need for a European federation, though without direct British involvement: 'We have our own dream and our own task', he wrote in 1930, 'We are with Europe but not of it. We are linked but not compromised. We are interested and associated but not absorbed'.[56]

'Continental' supporters had no doubts about their European credentials. Herriot, then Prime Minister of France, spoke openly of the need for a 'United States of Europe' in 1924, subsequently producing a book with that title which argued that a united Europe was an inevitable outcome of economic processes that were 'silently creating a new world'.[57] Briand was equally enthusiastic and became the Honorary President of the PEU. Renowned, some would say notorious, for his florid oratory (his speech welcoming Germany into the League of Nations in 1926 is said to have reduced even the simultaneous translators to tears), Briand formally proposed a 'confederal' scheme for European union in a speech in Geneva on 5 September 1929. Nine months later, on 17 May 1930, he circulated a memorandum, with the full weight of the French Foreign Ministry behind it, to 26 European governments, including Britain but excluding Russia and Turkey. This called for a formal European economic and political union, a federal Europe 'in a single bound' with its own permanent secretariat and council of elected representatives.[58] The document, though prolix, was remarkably vague on the crucial economic questions, as Briand candidly admitted. Economics, he believed, 'were at the mercy of politics in Europe'. His plan was based on 'the general subordination of the economic problem to the political'.[59]

Briand's initiative was premature and was to founder completely after his fall from power.[60] The responses to his memorandum were polite but cool. Mussolini was unimpressed and although there was widespread support in Germany for a European customs union to stimulate trade (an idea championed by the PEU and which the League of Nations had attempted to promote through the 1927 World Economic Conference), these economic considerations were precisely what Briand chose to ignore.[61] The mandarins in the British Foreign Office were baffled by the 'vague and puzzling idealism' of Briand's document. Some suspected a sinister Gallic plot to create a European secretariat which France could then dominate.[62] The influence of Coudenhove-Kalergi on Briand's thinking went down especially badly with Sir William Tyrell, the Permanent Under-Secretary. The idea was 'fantastic', he blustered, 'thoroughly impractical'.[63] This is not to say that British political opinion as a whole was antithetical to European integration. Churchill's interest has already been noted and Britain's business community was also enthusiastic about a European customs union.[64]

Radical regionalism

A more radical democratic perspective on European unity rejected the pragmatism of the PEU, particularly its acceptance of the existing nation-states.

A united, democratic and peaceful Europe could never emerge, it was argued, while traditional, rigidly territorial nation-states persisted. Such states were defensible spaces, designed partly on the basis of their ability to wage war. They were therefore antithetical to genuine internationalism and the idea of a united Europe. European unity demanded a new political geography of smaller, regional units which could then coalesce to establish a federation based on regions rather than states. Whereas the PEU sought to overcome the antipathies of nation-states from above, the regionalist argument sought to do so from below. This viewpoint represents a direct continuation of the regionalist critique of the nation-state, discussed in Chapter 2.

Regionalist ideas on European federation were especially strong in Italy and Spain before being snuffed out by Mussolini and Franco in the mid-1920s and mid-1930s, respectively.[65] Similar perspectives were also widely debated in the United Kingdom, the country whose diplomats had been so unimpressed by Briand's PEU-influenced proposal. This is not to say that regionalist arguments carried the day in inter-war Britain, one of the more centralised European states despite a constitution built on the idea of a union between distinctive nations. Indeed, one distinguished historian has described British inter-war regionalism as 'the dog that never barked'.[66] But the secession of the Irish Free State in 1922 sustained an intense debate about the nature of state sovereignty. The highly regional impact of the depression on the UK economy during the early 1930s also accentuated calls for political decision-making to be devolved downwards, part of a wider renaissance in urban and regional planning.[67] These factors were connected to what David Livingstone calls the 'regionalising ritual' of political and academic debate in inter-war Britain, a 'ritual' in which the names discussed in Chapter 2, notably Patrick Geddes, loomed large.[68] A succinct statement of the regionalist view was provided by H.J. Fleure in a pamphlet on the new political landscape of Europe. Here he revealed a regionalist's frustration that, despite the new political geography, the old, discredited geopolitical ideas had been retained:

> Beyond vague clauses in the Convenant of the League of Nations, there has been little thought of avoidance of future wars by the encouragement of widespread aspirations, especially of the humbler folk, towards a growth of European unity. ... The state, as we have known it, has been provocative of war by the exclusiveness on which it is based ... it needs to be tempered by the creation, not of a super-state, but of the inter-state Court of Justice which many small nations, often the real trustees of civilization, so ardently desire. The unfortunate, but characteristic, opposition of the dominant states, whose power would be greatly reduced were it to be instituted as a reality, is delaying if not defeating this hope of peace. The dominant states have so framed the economic sides of the treaties as to make the rise of unity difficult, and the recurrence of war almost inevitable.[69]

The distinguished constitutional lawyer, academic, politician and diplomat, James Bryce (later Lord Bryce) was an important influence on this argument. Bryce's view of Europe extended that of his fellow Scottish lawyer, James Lorimer (see Chapter 1), and reflected his historical interest in the Holy Roman Empire and his professional concern with US constitutional law. Bryce rejected territoriality as the basis of popular sovereignty. Instead, he suggested a system in which sovereignty was shared between interconnected local, regional and national levels of government, all overseen by an *imperium*, a supreme peace-keeping agency which would have both secular and religious figureheads.[70] Similar ideas were developed by Philip Kerr, later Lord Lothian, Lloyd George's private secretary during the war. Kerr believed nation-states should no longer exercise absolute sovereignty within distinct territorial units. Rather, sovereignty should be 'pooled', downwards toward regional assemblies and upwards towards wider federations, including Europe. These would, in time, spawn a global confederation.[71]

The prospect of the old nation-states withering away under pressure from above and below, hastening the emergence of European and ultimately world unity, was immensely attractive to many on the British left, including the anti-war campaigners H.N. Brailsford, Bertrand Russell and Harold Laski as well as other, less overtly pacifist left-of-centre intellectuals such as H.G. Wells and G.D.H. Cole.[72] It should be emphasised, however, that many British socialists, particularly in the trade union movement, suspected any project of European federation on the grounds that such an organisation would inevitably reflect capitalist business interests, whatever the original motives which lay behind it.

Those Britons who yearned for some form of international government were genuinely ambivalent about their best course of action. Should a British campaign focus on Europe or the Empire? Indeed, were these two alternatives mutually exclusive? Churchill, as we have seen, still opted for the 'open seas' and the imperial frontier while welcoming top-down European unity. Other politicians questioned Britain's capacity to operate independently of Europe. Neville Chamberlain seemed ambivalent about Britain's destiny in 1929: 'We, in these islands, cannot stand by ourselves alone. If we do not think imperially, we shall have to think continentally'.[73] Five years later, after the rise of Hitler, Stanley Baldwin expressed the dominant view that Britain no longer had any choice but to play its part in the European struggle:

> Let us never forget this: since the day of the air, the old frontiers are gone. When you think of the defence of England you no longer think of the chalk cliffs of Dover; you think of the Rhine. That is where our frontier lies.[74]

In these circumstances, British regionalist debates tended to encompass both Europe and the Empire. The writings of Lionel Curtis, the colonial historian and prominent member of the Oxford Round Table group of colonial reformers, were especially intriguing. Curtis was a committed federalist of conservative leanings, convictions born of his experience in helping to draft the constitution for the Union of South Africa. In his remarkable 1935 text, *Civitas Dei*, Curtis looked forward to a world of great federations, all operating under an agreed system of global governance.[75] C.B. Fawcett also argued for a new, less obviously colonial empire, a trading commonwealth. Despite its geographical dispersal and the dangers of imperial overstretch, Fawcett hoped that the Empire might yet emerge as a pan-region operating in part as an outlet for an otherwise self-sufficient European economic space. Imperial reform, the catch-phrase of Joseph Chamberlain's earlier campaign which had so influenced Mackinder, was here invoked in pursuit of inter-regional trade within a wider pan-region, the model for a future 'world commonwealth'.[76]

The Weimar critiques: the geopolitics of resentment

In Weimar Germany liberal appeals for European unity failed to convince those who resented both the treatment their country had received in 1919 and its subsequent humiliation. The rise of a distinctive and self-conscious 'school' of German *Geopolitik* during the 1920s reflected this anger and produced some very different visions of Europe. Just as wartime geographical research in France sought to legitimise French claims to the 'lost' territories of Alsace and Lorraine by drawing on the illustrious scholarly traditions of French regional geography, so patriotic German geographers attempted to demonstrate the injustice of the Treaty of Versailles by developing post-war critiques of the existing order based on the still prevalent geopolitical theories of Ratzel and Kjellén. Two distinctive but overlapping critiques developed.

The first, linked to what George Mosse has called *völkisch* German nationalism, sought to challenge the Treaty of Versailles by exposing the manifest hypocrisy of the self-determination argument which had allegedly underpinned the creation of the new central European states.[77] Germany's right to self-determination had clearly been ignored in 1919, making a mockery of this supposedly universal principle. The Berlin geographer Albrecht Penck, whom we met in Chapter 2, was a vocal proponent of this argument. Penck and his co-workers in Berlin and at an official Leipzig-based research organisation, the *Stiftung für deutsche Volks- und Kulturbodenforschung* (Foundation for Research on the German People and Cultural Region) sought to delimit the 'natural' frontiers of Germany. A series of striking ethnic maps and atlases were produced, together with a stream of books and articles on the 'stranded' German enclaves beyond the restrictive limits of the new Germany, Penck himself writing extensively on the German population living under Polish rule in the 'Danzig corridor'.

The objective was to highlight the yawning discrepancy between Germany's 'natural' limits and the territorially deprived Weimar Republic. These limits were pushed to their furthest possible extent. Penck distinguished between *Volksboden*, the zone occupied by German-speaking peoples stretching far beyond the Weimar borders, and *Kulturboden*, an even more extensive area where German *Kultur* was predominant, as measured by a complex range of cultural practices, building styles and landscape features. These two zones featured in dozens of inter-war maps and atlases (Figure 3.7).[78]

Penck's contribution was part of a wider campaign not only to challenge the legitimacy of the Treaty of Versailles but to stress the urgent need to

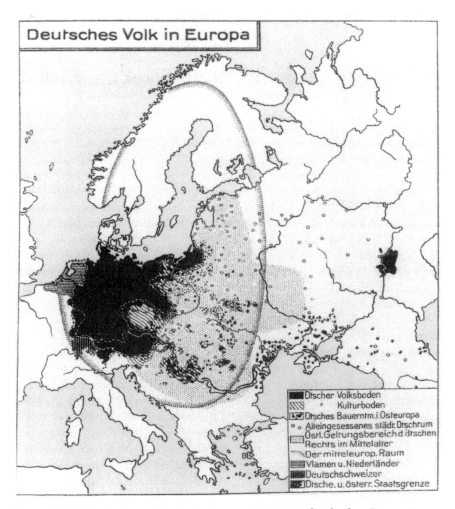

Figure 3.7 *Volksboden* and *Kulturboden* in a Weimar school atlas. Source: Braun and Ziegfeld (1929–30, vol. III, *Die Neuzeit*, 43).

unify all Germans in a new geopolitical arrangement, the only long-term solution to Europe's enduring problems. Precisely what form this arrangement would take stimulated some interesting debate. Penck's approach was strongly influenced by the Ratzelian concept of *Volks- und Kulturboden*, the idea that a 'natural' and measurable affinity existed between people, culture and soil. As we noted in Chapter 2, Ratzel and Kjellén developed their organic theory of the state as a Hegelian critique of French bourgeois republicanism, the legacy of 1789. Where Ratzel and Kjellén emphasised the interdependence of state, people and soil, the rational 'bourgeois' view adopted a much narrower, legal definition of the state as a discrete, overarching institution whose relationship with its citizens was mediated through technical rights and responsibilities. Penck and the *völkisch* German nationalists argued that the Treaty of Versailles represented the triumph of the 'bourgeois' nation-state. The conventional nation-state had been accepted as the only possible building-block for the new Europe and imposed everywhere. Yet such an arrangement was inappropriate for Germany and much of central Europe. Not only were the borders of the Weimar state wildly at odds with the racial map; they reflected mistaken assumptions about how the nation-state should operate in central Europe. The German and central European peoples required a more flexible geopolitical system, more suited to the predominantly Hegelian culture of the region.

This subtle argument frequently collapsed into a straightforward demand for a reconstituted German *Reich* encompassing as many ethnic Germans as possible. But it also inspired renewed calls for a central European customs union, a *Mitteleuropa*, transcending the limiting national borders imposed in 1919 and also holding together the 'two Europes' which Alfred Weber, the author of the famous theory of industrial location, detected in the 1920s. The agricultural east and industrialised west were obviously complementary, claimed Weber, but were instead growing apart. This tendency, which would spell the end of Europe, needed to be reversed by co-ordinated action, led by Germany.[79] The work of M.H. Boehm, director of the Berlin-based *Institut für Grenz- und Auslandsstudien* (Institute for Border and Foreign Studies), was also important in this respect. He sought to map a German economic space linked to the German *Volksboden* (or, as he preferred, *Volkssiedelboden*). The wide dispersal of the German people, he argued, and their economic influence across central Europe demanded an inclusive economic space across the whole of *Mitteleuropa* which would logically arise from two different trading blocks, German in the north and Danubian in the south, the latter linking Austria, Czechoslovakia and Hungary.[80]

The second Weimar critique, though equally influenced by the Hegelian ideas of Ratzel and Kjellén, had less obvious *völkisch* connotations. The dominant influence here, and the leading figure in the academic geopolitical movement under both the Weimar Republic and the Nazis, was Karl Haushofer, a retired Bavarian general who became Professor of Geography

at the University of Munich in 1921.[81] Haushofer's geopolitical writings
were aimed partly at the intellectual, academic community and partly at a
more popular audience. Like Penck, Haushofer also sought to demonstrate
the injustice of the Treaty of Versailles and established a monthly review as
a forum for discussion, *Zeitschrift für Geopolitik*, which he edited for over
20 years from its establishment in 1923. This was a brash exercise in geo-
propaganda, with short, punchy articles and strikingly suggestive thematic
maps often in stark black and white, the articles conveying the idea of a
threat to Germany or Europe, the maps suggesting innocence or vulnerabil-
ity. The maps emphasised the strategic and economic weakness of the
diminished and fragmented Weimar Republic and became increasingly
elaborate. They were also widely copied in German school textbooks and
atlases (Figure 3.8).[82]

Haushofer's critique of the Treaty of Versailles culminated in a three-vol-
ume, multi-author work published in the early 1930s under the title *Macht
und Erde* (Power and the Earth). Here Haushofer spoke of Germany's
'imprisonment', the harsh deprivation of its 'vital space'. A mass of statistics
were produced on the ratios of population to available space in different
countries: if the French ratio was 1.0, claimed Haushofer, the British imperial figure was 3.5, the Russian figure 2.5 and the German index a mere 0.6.
Fewer than 60 per cent of Europe's 100 million Germans had lived within
the 1914 borders, he complained, and the butchered Weimar state was even
more restricted.[83]

But Haushofer's vision went far beyond a narrow concern with the limitations of the Treaty of Versailles. Like Mackinder, he was intrigued by
global shifts in the economic and geopolitical order and by the 'imminent'
appearance of self-sufficient pan-regions. His imaginative use of suggestive
and symbolic cartography, often at continental scale, reflected this expansive brand of European geopolitics. Although critical of Britain, Haushofer
admired Mackinder's logic (if not his conclusions): 'Never have I seen anything greater', he wrote of Mackinder's 'geographical pivot' paper of 1904,
'than those few pages of geopolitical masterwork'.[84] Whereas Mackinder
saw the development of a Eurasian 'heartland' as a direct challenge to the
British Empire, 'the real Europe' and capitalist liberal democracy,
Haushofer welcomed this development because he believed such an arrangement would pose a serious challenge to the slow 'spiritual death' represented by the morally debilitating capitalism which had taken hold in the
USA, Britain and France. One of Haushofer's main interests was Japan,
where he lived from 1909 to 1912, and he enthusiastically endorsed the idea
of a 'transcontinental bloc', a great Eurasian pan-region limited by the
Rhine in the west and the Yangste in the east and based on a tripartite
alliance between Germany, Russia and Japan (possibly even China and
India).[85] The traditional divide between Europe and Asia would become
increasingly meaningless, he argued, as a 'natural' land empire linking these
historically distinctive zones was poised to develop.

The presence of old-style European imperialism in Asia was a serious obstacle to this geopolitical transition. The imperial domination of the 'Indo-Pacific space' by Britain and France imposed economically unrealistic and geographically unsustainable colonial ties across thousands of miles of ocean, corrupting the more obvious relationships. The European imperial powers needed to confront the harsh 'geographical realities' described by Mackinder in 1919 and accept that their distant colonies in the Pacific, South Asia and the Far East would soon shake off their colonial status and reorientate themselves with respect to a new Eurasian system.[86] This would mark the first step from the east towards a transcontinental bloc and should be matched by a comparable development in the west: the establishment of a German central Europe, a new, self-sufficient *Mitteleuropa* which should likewise free itself from the corrosive materialism of 'western' capitalism. 'The struggle for the liberation of India and China from foreign domination and foreign capitalist pressure', he wrote in 1924, 'meets with the dreams of Central Europe'.[87]

Haushofer's vision of Europe sublimated into a new Eurasian realm rested on his belief in the natural spiritual affinity between Germany and Russia, two peoples who together could save the world from the pernicious influence of twentieth-century capitalism. Whereas liberal critics such as Coudenhove-Kalergi saw Russia as a dangerous and alien state because it had embraced communism, Haushofer wrote of a transcendent Russian soul which operated independently of any governing political philosophy. Whether led by communists or a Tsar, the Russian people were ultimately unchanging. Russia's fundamental geopolitical instincts and aspirations were, he argued, constant. Notwithstanding the upheavals of the Revolution, Russia's geographical position determined its status as the natural conduit between Europe and Asia.[88] Interestingly, Mackinder had articulated precisely this view some 13 years before the October Revolution. 'Nor is it likely', he wrote in a 1904 paper, 'that any possible social revolution will alter her [Russia's] essential relations to the great geographical limits of her existence'.[89] Haushofer's positive view of Russia, and his belief in a natural alliance with Germany, were influenced by Oskar von Niedermayer, a friend and fellow Bavarian geopolitical theorist.[90] Von Niedermayer was Germany's principal military attaché in Moscow during the 1920s where he forged close links with senior officers in the Red Army before returning to take up a chair of geography at Berlin in 1932. Von Niedermayer, who was known as 'Major Neumann' during his Moscow period, had always advocated German expansion to both east and south (*Drang nach Osten* and *Drang nach Süden*), and now believed that this should be combined with a renewed southern expansion of Russian influence to challenge British colonial rule in south central Asia, a strategy he actively encouraged during his Russian sojourn.

Racial theorising was largely absent from Haushofer's geopolitics during the Weimar period. Anti-semitism, so central to Nazi ideology, had no

(a)

XXXVII. Kriegsende und Friedensschlüsse.

(b)

XXXVIII. Kriegsende und Friedensschlüsse.

Figure 3.8 The cartography of resentment: Weimar school atlas (a) 'Germany's Mutilation', (b) 'Germany's Enslavement' and (c) 'Germany's Imprisonment'. Source: Braun and Ziegfeld (1929–30, vol. III, *Die Neuzeit*, 37–8, 40). Map (a) dwells on the country's loss of territory and population, map (b) laments the country's economic humiliation and map (c) focuses on the overwhelming military power of the surrounding states.

discernible influence on his writings (indeed his wife, Martha, was half-Jewish). Haushofer's primary concerns were spatial and territorial rather than racial or ethnic. His vision of a central European federation (with Germany as the dominant power) was multi-ethnic as was his wider view of a Eurasian pan-region which he saw as a transcultural and transracial zone, based on alliances between broadly equal partners. Haushofer's writings under the Weimar Republic represented a direct extension of the Ratzelian tradition, though the style and details of his arguments reflected the new conditions. Taken together, the work of Haushofer and Penck demonstrate the continuities in German geopolitical thinking from the late nineteenth-century imperial era into the Weimar Republic.[91]

Haushofer's proselytising was roundly condemned in other countries. In the USA, Isaiah Bowman went to great lengths to distinguish his 'neutral' and 'objective' political geography from Haushofer's 'crudely partisan' geopolitics.[92] In France, the students of Vidal de la Blache, who still dominated French geography, were especially hostile. André Siegfried claimed Haushofer's writings induced nausea, and Yves Goblet spoke of German geopolitical 'spagyrics'.[93] Others spent their time refuting the accepted tenets of German geopolitical writings. Emmanuel de Martonne produced a two-volume work on central Europe in which he rejected the idea that this region could ever form a coherent geopolitical block.[94] Joseph Aulneau went further, seeing the idea of *Mitteleuropa* as 'an evil fiction . . . existing only in the imagination of conquerors and writers'.[95] Other French geographers sought to challenge Haushofer on his own geopolitical terrain. Jacques Ancel produced a book on geopolitics in 1936 and a review of frontiers and boundaries two years later, a direct response to Haushofer's 1927 work on the same topic.[96] Some of Ancel's arguments were even more disturbing than Haushofer's, though he concluded by arguing that Europe's international frontiers would ultimately collapse: 'The walls of these Jerichos', he claimed, 'will fall at the sound of the trumpets which will awaken the bound and sleeping nations'.[97]

Nuova civiltà: Italian fascism and the idea of Europe

There was a short but significant step from the Weimar critiques of capitalist, bourgeois Europe to a more radical and reactionary attack on the very idea that Europe should seek to preserve liberal democracy. Some believed this was a bankrupt ideology, the legacy of a corrupt, bourgeois age which had brought Europe to its knees. If Europe was to survive it needed a new ideology for a new century. Fascism was the most disastrous alternative to liberal democracy to emerge from the maelstrom of World War 1, though its roots lay in the more militant and authoritarian strands of late nineteenth-century European nationalism.[98]

Fascism is extremely difficult to define, partly because it is a fundamentally irrational ideology, deriving its force from exploiting fears, anxieties and phobias which defy logical analysis. For this reason, Ernst Nolte has described fascism as an oppositional ideology, wedded only to a series of negatives – anti-liberal, anti-communist, anti-conservative, anti-consensus.[99] The fact that many regimes now commonly interpreted as 'fascist' rejected this label (including the Nazis) and developed their own, distinctive political programmes and objectives makes the definitional problem especially acute. Some have argued that using the term 'fascist' to bracket together Italy, Germany, Spain and other authoritarian regimes of the 1920s, 1930s and 1940s erases their many differences. There was not one fascism, it is claimed, but many; so many, in fact, that the term itself becomes problematic.[100]

Jeffrey Herf suggests that fascism can usefully be seen as 'reactionary modernism', a complex belief system which drew on at least two contradictory ideals.[101] First, there was a predilection for a romantic mythology about a glorious past which many fascist leaders claimed to embody. The belief in former glory and present decay was a common element in fascist rhetoric. Most dictators claimed to be driven by a desire to re-establish their nation's historic role, its lost pride and dignity.[102] Yet Europe's inter-war fascist movements were also inspired by a triumphantly modernist faith in the power of science, technology and rational planning, manifested most obviously in the fascist obsession in Italy and Nazi Germany with speed, energy and power. Whereas the former component of fascist thinking looked back, often to a bucolic age of harmony between a nation's people and its natural environment, the latter looked forward to a new scientific utopia when nature was thoroughly conquered and controlled.[103]

Despite these complexities and variations, fascism was everywhere, as Roger Griffin notes, 'a form of populist ultra-nationalism', a twentieth-century corruption of eighteenth-century and nineteenth-century debates about nations, cultures, peoples and races fused with crushingly modern forms of militarism and authoritarianism.[104] It is also important to recognise that, despite their ultra-nationalist tone, the various forms of fascism which flourished in Europe between the wars were also pan-European. They were, in part, based on wider visions of the entire continent.[105]

Italy, the world's first overtly fascist state, was the obvious source of alternative views on the European geopolitical order. Mussolini's Blackshirts seized control of Rome in a 'bloodless' coup in October 1922, fuelled by the rising tide of resentment at the chronic instability and corruption of the liberal Italian republic and by the enduring bitterness about the nation's treatment at the hands of its wartime allies in 1919.[106] Mussolini's fascist programme, developed gradually through the 1920s, was primarily concerned with domestic and then non-European colonial matters. But in his warped imagination, fascist Italy was ancient Rome reborn in the modern age. It was the kernel of European culture, civilisation and law. Mussolini's grandiose plans to transform Rome and its surrounding region into a fitting imperial arena as well as a modern European capital were based on this vainglorious rationale. The new Italy must be prepared confidently to reassert its historic role at the heart of a new Europe.

The European rhetoric of Italian fascism was sufficiently encouraging for Coudenhove-Kalergi to request an audience with *Il Duce* in May 1933, five months after Hitler's chancellorship began. This was a desperate and misjudged attempt to convince the dictator that Italy should lead a union of southern European 'Latin' peoples, a confederation of states which would be a stepping stone towards a new Europe. Not surprisingly, the meeting did not go well, particularly when the Austrian began to lecture the bewildered Italian on Nietzsche's 'Europeanism'.[107] Liberal pan-Europeanism based on the idea of equality between all nations, big and small, was anathema to

Mussolini's moral and ideological convictions, as his dismissal of Briand's memorandum three years earlier had demonstrated.

Mussolini's vision of a reborn, integrated Europe rested on two preconditions. First, he believed that a degree of unity could only arise if the 'great' powers agreed to share out, in a 'just' and 'rational' manner, their colonial responsibilities both with respect to the 'smaller peoples' of Europe, whose size and/or civilisation denied them the right to nationhood, and with respect to the non-European colonial world. While the territorial injustices introduced in 1919 prevailed, both in Europe and around the world, European unity could never arise. Second, Mussolini argued that European unity would need to be forged in a common struggle against the dual threat of communism (the USSR) and unfettered capitalism (the USA). These aversions were evident in liberal visions of a united Europe, but in Mussolini's mind they acquired almost phobic intensity. The USA, to which so many Italians had emigrated over the preceding half century, was represented as a serious moral and spiritual threat to European culture and civilisation; it was the heartland of the spirit-sapping materialism of twentieth-century capitalism.[108] The USSR represented an equal but opposite evil which also threatened to destroy Europe's historic mission. Europe's crusade would only be victorious if fascist-style authoritarianism was embraced across the entire continent, a *nuova civiltà* of 'universal fascism' personified by the macho ideal of 'Fascist Man'.[109]

The precise nature of the *nuova civiltà* was the subject of lively debate. Writers such as Edgardo Sulius, Giovanni Castellani, Ezio Maria Gray and Italy's 'poet of fascism', Gabriele D'Annunzio, emphasised the need to preserve the cultural integrity of 'old' Europe from alien, modern influences. Their fascism was essentially conservative; it was, they believed, the only force standing between civilisation and barbarism. A younger fascist group, with links to the Italian futurist movement, represented by Filippo Marinetti and Italo Balbo, rejected the past altogether and sought to create a new technological utopia across Europe.[110]

By the late 1930s, fascist debates about Italy's role in Europe and the wider world had spawned a formal geopolitical movement with its own journal, *Geopolitica*, established in January 1939. The Italian geopolitical movement was strongly associated with the port city of Trieste, on the eastern flank of the Gulf of Venice, which was ceded to Italy from the former Austria–Hungary in 1919. Like French inter-war governments which built up the University of Strasbourg as a proselytising centre from which French civilisation would suffuse through the reclaimed provinces of Alsace and Lorraine, the Italian government transformed the University of Trieste into a centre of Italian learning for similar reasons. But whereas Strasbourg became the spiritual home of Europe's finest school of history, Trieste spawned an ignoble tradition of populist geopolitics devised by strutting, self-important Blackshirts such as Giorgio Roletto and his Germanophile student Ernesto Massi.[111]

The fascist geopolitical imagination in inter-war Italy exhibits many superficial similarities with Haushofer's German school under the Weimar Republic, both in style and content. *Geopolitica* emulated the format of Haushofer's *Zeitschrift für Geopolitik* and there were regular contacts and exchanges between members of the two national 'schools'. Most articles were written in the same style and were accompanied by similar suggestive maps. There were, however, several important differences. To some extent, these reflected the different interests of the two countries. *Geopolitica* dwelt on Italy's territorial claims in Africa, the Balkans and the Middle East, demands which had been so contemptuously denied in 1919.[112] But there were more substantive differences which grew wider over time. Italian fascism saw itself as a transformative, invigorating ideology to which all European classes and races could subscribe. Racism, particularly the anti-semitism that became so central to Nazism, was far less prominent in Italian geopolitical debates on Europe. Racial theories were by no means absent, of course, as the discriminatory racial laws of 1938–39 and the associated establishment of Italian journals such as *La Difesa della Razza* (1938), *Il diritto razziste* (1939) and *Razza e civiltà* (1940) make clear. But for much of the preceding 20 years, and even during the war itself, Italian geopolitical debate had remarkably little to say about race and was more concerned with the ideological affinity of different European nation-states for fascist-style government. As we shall see, this may partly explain why a much lower proportion of Italy's Jewish population were murdered during the war than might have been expected in Germany's principal European ally.[113]

A new order: Nazism and the idea of Europe

The rise of the Nazis marked a significant departure in German and European geopolitical thinking. There is a huge literature on the Nazi seizure of power.[114] Suffice it to say that Germany's economic collapse during the early 1920s fuelled political extremism of all kinds, including the infant Nazi party (the *Nationalsozialistische Deutsche Arbeiter Partei* or NSDAP), the National Socialist German Workers' Party, established by an unknown corporal, Adolf Hitler, and an equally unimpressive assortment of disaffected ex-soldiers in February 1920.

German inflation rates had risen from 9 marks to the US dollar in January 1919 to 18 000 in January 1923 (when French troops occupied the Ruhr). Thereafter, the currency went into free fall; by November 1923 (when Hitler organised his abortive *Putsch* against the provincial government in Munich) the exchange rate stood at four million marks to the dollar.[115] Economic conditions stabilised thereafter, halting the rise of political extremism through the mid-1920s. While the NSDAP polled 6.5 per cent of the vote (mainly in the south) on 4 May 1924 and returned 32 *Reichstag* deputies, support had fallen to just 2.6 per (12 deputies) four years later.[116] Thereafter, economic

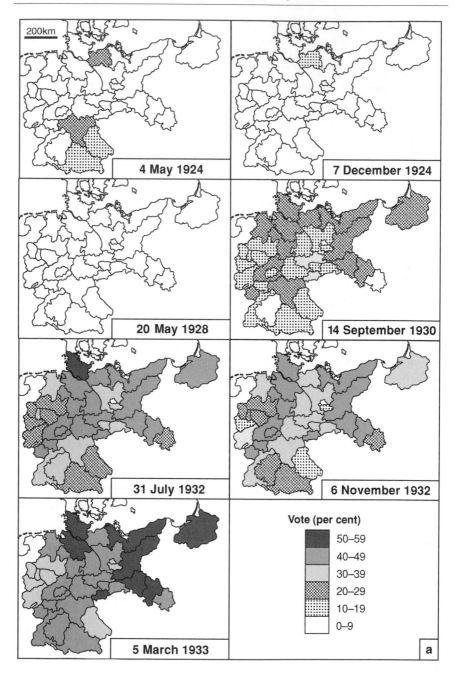

Figure 3.9 The Nazi vote. Source: after Freeman (1995, 27, 39, 35–7, 43), Bullock (1991, 240, 1082–3).

and political circumstances shifted inexorably towards the Nazis who skilfully exploited the mounting crisis, not least through violence and intimidation.[117] By the autumn of 1930, the NSDAP was the second largest party, with 107 *Reichstag* deputies compared with 143 socialists from the SPD (*Sozialdemokratische Partei Deutschlands*). Over 18 per cent of the German population (6.5 million people) voted NSDAP.[118] Two years later, in the elections of 31 July 1932, almost 40 per cent of the electorate (14 million people) voted for the Nazis. With 230 *Reichstag* deputies, the NSDAP was easily the largest party, outnumbering the socialist SPD and the KPD (*Kommunistische Partei Deutschlands*) deputies added together. By March 1933, the NSDAP's dominance in the *Reichstag* was unassailable: 44 per cent (17 million people) returned 288 Nazi deputies (see Figures 3.9 and 3.10).[119] Hitler also stood against General Paul von Hindenburg, the reactionary wartime military leader, in a two-round presidential election in the spring of 1932, polling 30 and 37 per cent of votes respectively, with his highest support registering in the northern rural areas of East Prussia and Schleswig–Holstein. Hindenburg retained the presidency but Hitler's politi-

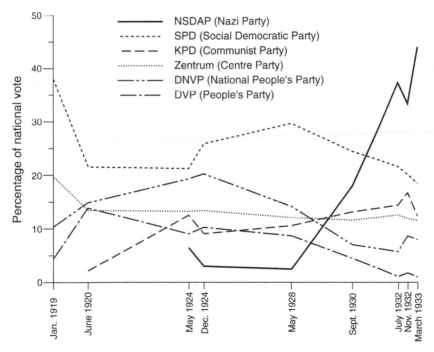

Figure 3.10 Votes cast at German elections. Note: NSDAP, *Nationalsozialistische Deutsche Arbeiter Partei*; SPD, *Sozialdemokratische Partei Deutschlands*; KPD, *Kommunistische Partei Deutschlands*; DNVP, *Deutschnationale Volkespartei*; DVP, *Deutsche Volkspartei*.

cal momentum carried him to the chancellorship, the principal executive position, in early 1933 amidst escalating violence on the streets, and complex, behind-the-scenes political 'horse-trading'.

Hitler's grip on power quickly tightened, particularly after his brutal execution of would-be rivals within the NSDAP in the summer of 1934.[120] Rigid controls over the press and the educational system, backed up by the ever-present threat of state terrorism, forced those academics who chose to stay in Germany to modify their message. Many geopolitical theorists actively supported the Nazis. Those who did not tended to emphasise the aspects of their research which struck a chord with Nazi thinking. This was no easy task, for Nazism was neither coherent nor consistent. As a dictatorship, official thinking reflected the Führer's wishes, but these were generally vague and were determined as much by circumstance and opportunity as a carefully devised 'blueprint' for Germany or the rest of Europe.[121]

There were, however, continuities and connections between Weimar geopolitics and Nazi thinking. The Ratzelian critique of the 'bourgeois' nation-state as an alien, non-Germanic imposition was accepted by many Nazi theorists.[122] Haushofer could also boast influential supporters close to the centre of Nazi power. Rudolf Hess, Hitler's second-in-command until his mysterious flight to Scotland in 1941, allegedly to broker peace with Britain, was a close personal friend. Joachim von Ribbentrop, the Nazi foreign minister from 1938, also drew on Haushofer's ideas (see below).

Hitler, however, had little time for Weimar geopolitics. This was largely because Hitler's National Socialism was as much a racialist as a nationalist ideology. The desire to ensure Germany's racial purity determined Hitler's geopolitical objectives. In the Nazi view, the failure of Haushofer (and the Italian geopolitical theorists) to position race at the heart of their analysis compromised their entire project. The 'limited' geopolitical objectives of Penck and the *völkisch* nationalists were likewise unacceptably modest and failed to emphasise the 'natural' superiority of the German race over other peoples.[123] The idea that a new, inclusive German *Reich* might simply exist alongside 'lesser' European races or even act as a benign focus for a cosmopolitan, multi-ethnic *Mitteleuropa* were almost laughably at odds with Hitler's view.[124] This version of *Mitteleuropa* recalled the detested bourgeois liberalism of Coudenhove-Kalergi, 'everyone's bastard' as Hitler called him in 1928.[125] Cosmopolitanism implied degeneration. It was an evil, quintessentially Jewish trait, a deliberate conspiracy to accelerate the decay and collapse of Europe. In the words of Erhard Wetzel, later the author of the *Generalplan Ost* – the wartime blueprint for a Nazi-dominated Eastern Europe drawn up for Heinrich Himmler's Schutzstaffel (SS) – 'the greatest misfortune which could threaten the German *Volk* would be the triumph of the pan-European racial idea, which could only have as a consequence a great European racial swamp'.[126]

This brings us to the core of Hitler's racial ideology, the central and most consistent element of his doctrine: anti-semitism. The frightful consequences of Nazi anti-semitism during World War 2 will be discussed later in this chapter, but it is important briefly to examine how the irrational fear and hatred of Jews became a key component in Nazi geopolitics before World War 2. The persecution of Jews has been a perennial feature of Europe's history, the subject of many exhaustive analyses.[127] European anti-semitism derives partly from Christian intolerance of another faith, but its persistence, indeed its intensification, into the secular twentieth century is directly connected to the intensely territorial nature of European cultural and political identities. Europe's Jews could (and should) have been viewed as emblematic of a future united European population. Although the majority were poor and rural, those who became the subject of the most vehement hatred were urban, sophisticated and cosmopolitan. This was a community which consciously rejected the narrow tyrannies of place. But in a Europe which defined itself by the desire for rootedness, for a sense of geographical belonging, the figure of the diasporic 'wandering Jew' appeared as a threat to this supposedly universal, 'natural' condition. Jews were alien to the European perspective; they lacked that sense of direct connection with place which defined what it meant to be European. Anti-semitism was depressingly common all over Europe but it has been argued, controversially, that the Hegelian, organic view of the relationship between soil, people and state advanced by German intellectuals created an intellectual climate especially conducive to the more rabid forms of anti-semitism.[128] Hitler and his Nazi colleagues reflected and exploited a rich vein of popular anti-semitism, first in Germany and later across the whole of Europe.

In this context, it became possible to make the most absurd and contradictory claims about Jewish subversion of Germany, Europe, indeed the entire world. Jews were alleged to be simultaneously plotting an international capitalist conspiracy, centred on the USA, and an international communist conspiracy, based in the USSR. Secret Jewish societies had apparently engineered the French and Russian revolutions, World War 1 and the attempted German revolution of 1918–19. Anti-Jewish propaganda circulated freely around Europe during the 1920s, notably *The Protocols of the Elders of Zion*, a text purportedly derived from a secret Jewish sect, 'confirming' the existence of an international conspiracy. This was, in fact, a 1902 forgery concocted by the Tsarist secret police to justify their own persecution of Russian Jews.[129] A German translation of *The Protocols* appeared in January 1920 and by the time Hitler came to power there were 33 different editions in circulation.[130]

Alfred Rosenberg, the Nazi propagandist and editor of the NSDAP's newspaper, *Völkische Beobachter* (Racist Observer), was a key figure promoting anti-semitic mythology.[131] Rosenberg, who was also a passionate anti-Bolshevik, subsequently produced the unreadable *Der Mythus des 20 Jahrhunderts* (The Myth of the Twentieth Century) in 1930 which argued that Europe was in the grip of a titanic struggle between a decaying Roman Catholicism (represented by France), a rising Nordic righteousness (whose standard bearer was Germany) and a Jewish–Marxist conspiracy (led by the USSR). Germany was Europe's only potential saviour. Europe should so arrange itself as to ensure its racial and ideological purity. The western and southern flanks should be protected by France, Italy and Spain, leaving Germany, the soul of a reborn Europe, to resist the terrible dangers of the USSR. Like Coudenhove-Kalergi, Rosenberg saw Britain's role as a bridge between Europe and the Americas.[132]

Hitler saw a strong, racially determined German *Reich* built on National Socialist principles (*Kernstaat*) as the core of Europe. The inhabitants of this realm would be Aryan, a mythical racial category inherited by Nazi ideologues from nineteenth-century pseudo-science and energetically promoted as the racial category occupying the opposite end of the spectrum from the Jews.[133] The Aryan state, from which all physical and moral degeneracy would be eradicated by eugenic controls and strict laws regulating marriage, would then assume its 'naturally' dominant role, the home of a super race whose authority over 'lesser', mainly Slavic, peoples beyond the *Kernstaat* would be forever assured. The surrounding, non-Aryan people would have simple agrarian economies and would supply labour and basic foodstuffs to the *Reich*. There would be no place for Jews, nor gypsies, homosexuals and other 'undesirables'.[134]

Unlike Mussolini, who saw fascism as an 'exportable' ideology which might reshape Europe, Hitler insisted that Nazism was an exclusively German political code to which 'lesser' races could never aspire. 'It is in the nature of National Socialism to be exclusive', he once claimed:

> that is to say, it is concerned purely with the people of its own race and nation, whom it wishes to lead to a state of racial purity, unity and elevation, in accordance with their proper nature. It rejects elements contrary to the nature of its people, and has no desire to conquer or absorb them. . . . National socialism is not for export; it is not, like Communism, a plan for international subversion and subjugation.[135]

The desire to establish a racially pure Germany at the heart of a Europe organised according to a strict racial hierarchy determined Nazi territorial ambitions. The scale, direction and nature of these ambitions were neither clear nor consistent but were invariably more expansive and brutal than those suggested by Weimar geopolitical thinkers. *Mein Kampf*, the poisonous little book Hitler wrote during his spell in prison following the Munich *Putsch*, repeatedly refers to the need for German expansion into the east to create more *Lebensraum* for the *Reich*.[136] Whereas Haushofer stressed the affinity between Germany and Russia, and their common destiny in the east, Hitler saw the two nations as fundamentally antagonistic, both coveting the same spatial arena. Russia was, therefore, the ultimate 'Other' for most Nazis.[137] As the largest Slavic state it was racially degenerate; as the home of world communism (itself a Jewish conspiracy) it was also the principal ideological threat. Rosenberg articulated this fusion of racial and ideological venom in 1930: 'Bolshevism means the revolt of the mongol against the nordic form of culture. It is the desire of the steppe, it is the hatred of the nomad for a rooted personality, it means the attempt to wipe out Europe completely'.[138] Ultimately, a Germano–Russian war for eastern living space was inevitable.

Hitler's plans were described in detail in a rambling lecture before his puzzled military advisers on 5 November 1937. Notes taken on this occasion by Colonel Friedrich Hossbach suggest that race and space had become fused in Hitler's mind. Germany's renewal would be determined by a brutally simple formula: 'The aim of German policy is to make secure and preserve the racial community and enlarge it. It is therefore a question of space'.[139] *Lebensraum* was necessary to safeguard the race. Merely revising the Treaty of Versailles to re-establish the old Germany would not allow this; creating a German-dominated *Mitteleuropa* was also insufficient for this objective; colonial expansion beyond Europe (promoted by many, including the soon-to-be-dismissed economics minister, Hjalmar Schnacht) would similarly fail to deliver the desired results. The only viable arena for German living space was in the east, which Hitler, like many deluded European romantics, saw as the wellspring of the Aryan people. This sacred space would have to be won and held by force and would become the 'frontier' of the new *Reich*, guarded by a great *Ostwall*, the testing ground for future generations of German youth, a place of permanent, spiritually invigorating conflict with the alien, Slavic peoples beyond.[140]

Hitler's vision of an eastern German empire was staggeringly ambitious, stretching almost to the shores of the Caspian Sea. Poland, Czechoslovakia, the Baltic States, White Russia and the Ukraine were all firmly in his sights (Figure 3.11). Precisely what kind of regime was intended for these areas was never made clear but would certainly have involved the massive re-settlement of Russian and other Slavic peoples to allow space for German settlers, particularly in the Ukraine. Geopolitically, Russia would be pushed far into Asia. Other 'lesser' races would be tolerated outside the Reich but within Europe only as slave labour. Once this new order was achieved, the western European powers – Britain, France, Italy and Spain – would be free to develop their extra-European colonial interests, though these could never be as morally righteous as the forces which drove Germany to its eastern destiny.[141]

The Nazis ruled Germany for long enough before the outbreak of war to offer a fleeting but horrific glimpse of the kind of Europe they would have created had their desperate gamble paid off in 1939–41. From the moment he came to power Hitler sought to transform the landscape of Germany 'in his image'. The Nazi *autobahns*, a lasting legacy to Hitler's favourite engineer and planner Fritz Todt, had already formed a dense network of grey asphalt ribbons criss-crossing the country by 1939 and were in many respects a defining symbol of 'reactionary modernism'. Blending sinuously into the rolling German countryside, the *autobahns* were designed deliberately to

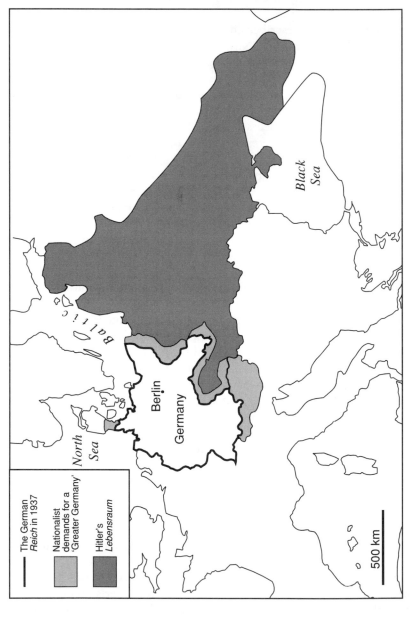

Figure 3.11 Hitler's Europe: the Hossbach Memorandum, 1937. Source: after Wright and Stafford (1987, 14).

appear as 'natural' landscape features whilst facilitating the conquest of nature by technology and the annihilation of space by time. The German road system would eventually spawn a great European network, binding together the new German *Reich* and its agrarian client states (Figure 3.12).[142]

Hitler was a frustrated architect, however, and the reconstruction of Germany's cities, the nodes in this great network, became his abiding passion. The plans, as ever, were tastelessly gargantuan. The initial idea to rebuild those cities of particular importance to the Nazis – Berlin (the capital city), Nuremberg (the spiritual home of Nazism), Munich (the NSDAP's birthplace) and Linz (the Führer's boyhood town) – were quickly revised upwards. Over 30 towns and cities were to be completely rebuilt, the most far-reaching programme of urban reconstruction ever contemplated.

Hitler's principal architectural advisers were Albert Speer and Hermann Geisler. Both men had worked on the German exhibits for the 1937 Paris World Fair, Speer designing the huge German pavilion which confronted the equally immense Soviet structure on the opposite side of the Seine.[143] Speer also organised the chilling Nazi rallies at Nuremberg in the mid-1930s and was then asked to rebuild the site of these gatherings, the centrepiece of which was to be the 400 000 seat 'German stadium'.[144] Speer was named *Generalbauinspektor* for the rebuilding of Berlin, the grandest of all Hitler's projects, in January 1937. By 1950, Hitler demanded, Berlin's central areas should be completely transformed and the city renamed Germania, a fitting capital for the '1000-year *Reich*'.

Speer's plan, which was still being formulated when war broke out in 1939, included a 5-kilometre grand processional avenue, the *Prachstrasse*, 120 metres wide with enormous official buildings on each side. Its southern end was to be marked by a triumphal arch which, at 120 metres tall, was to be three times the size of the Arc de Triomphe. Beyond, further to the south, would rise a colossal new railway station. The northern end of the central avenue was to be the site of a complex of massive government buildings linking the Brandenburg Gate and the *Reichstag* to the east, with Hitler's immense official residence to the west. Foreign dignitaries visiting the Führer would be obliged to walk 300 metres from its entrance to the official reception room via echoing marble halls, the objective being to humble them before the sheer power of National Socialism.[145] Looming above all these buildings would be the 300-metre dome of the great hall, the world's most imposing building, with room in its cavernous interior for over 150 000 people. No other city in the world would match the opulence and splendour of Germania, capital of Nazi Europe.[146]

Beyond the Revolution: the Bolshevik critique

As we have seen, the 1917 Bolshevik Revolution was followed by Russia's withdrawal from World War 1, the temporary transformation of eastern

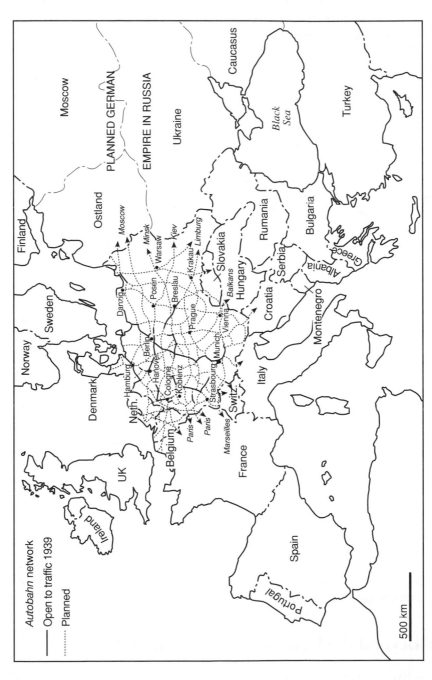

Figure 3.12 An empire of the *autobahn*: projections of German power. Note: place names in italics indicate the destination of the *autobahn*. Source: after Bullock (1991, 770–1); Freeman (1995, 70, 122, 171).

Europe's political geography following the Treaty of Brest–Litovsk, a bloody civil war to reconquer territories 'lost' under the terms of this accord and a series of international conflicts between the Red Army and enemy states. While Haushofer and others insisted the Revolution had not altered Russia's fundamental geopolitical challenges and problems, the triumph of Bolshevism provided a new power base for international communism and a spiritual home for Europe's Marxists. As so many schemes for the radical reorganisation of Europe had come from socialists and Marxists, many hoped that Bolshevik Russia would be a fruitful source of alternative thinking on Europe, the well-spring of a different geopolitical imagination.[147]

Marxism was predicated, however, on the idea of *world* revolution. Unfortunately for Marxist intellectuals, including those who came to power in Russia after 1917, their universalist claims left a host of strategic geopolitical questions unresolved. The first generation of Bolshevik leaders were hampered, therefore, by internal dissent about whether the revolution could be exported, and if so when, how and (most importantly) in which direction.[148] These strategic, geopolitical divisions developed into profound ideological disputes, particularly during the bruising battle to succeed Lenin during the mid-1920s.

Of the exiled revolutionaries who returned to Russia after 1917, Leon Trotsky was the most consistently internationalist and expansive. Revolution in one place was impossible, argued Trotsky. In order to survive counter-revolutionary pressures, the revolution had to be exported. Initially, Trotsky advanced an Asiatic policy for Bolshevik Russia. The new Russia should temporarily turn its back on 'the old Europe', he argued, relaunching itself as an Asian, Pacific power. At the same time, Bolshevism should foment anti-colonial revolution against the European imperial powers in Asia, thus hastening the final collapse of a teetering capitalist system in Europe.[149] Trotsky's Asiatic strategy was laid out in a secret memorandum of August 1919:

> the Red Army constitutes an incomparably more powerful force in the Asian terrain of world politics than in the European terrain. Here there opens up before us an undoubted possibility not merely of a lengthy wait to see how events develop in Europe, but of conducting activity in the Asia field. The road to India may prove ... to be more readily passable and shorter for us than the road to Soviet Hungary. ... The road to Paris and London lies via the towns of Afghanistan, the Punjab and Bengal.[150]

As this quotation indicates, Trotsky's interest in Asia was short-term and strategic. His more important objective was the creation of a communist Europe. Steeped in European culture, Trotsky believed Europe possessed a cultural unity which set it apart from the rest of the world. Its highly industrialised nature also gave it a privileged revolutionary position. In Trotsky's view, the industrial proletariat was the only true revolutionary class, their dependence on wage labour giving them a unique capacity for organised revolutionary action. Expansion into Asia was a means to this end rather than an end in itself. During his bitter power struggle with Stalin in the mid-1920s, Trotsky wrote increasingly of a 'Soviet United States of Europe' which would carry the battle for world communism against the USA, the epicentre of modern capitalism. 'Nothing compels us to remain in an atomized Europe', he wrote in 1926. 'It is precisely the revolutionary proletariat that can unify Europe, by transforming it into the proletarian United States of Europe'.[151]

This struck a chord with Trotsky's more doctrinaire Marxist followers, some of whom even welcomed a united, bourgeois and capitalist Europe to the west of Bolshevik Russia on the grounds that this would sustain the internationalisation of capital, a necessary precondition, according to Marx's predictions, for an international labour movement, itself a prerequisite for world revolution. It is also true that many Russian Bolsheviks, particularly those who had lived under the obfuscating Slavic mysticism of Tsarist Russia, were attracted to Marxism precisely because it was a European rather than an Asiatic philosophy, one which offered a rigorously scientific and materialist philosophy which would help to cement the connections between Russia and Europe, connections which an earlier generation of pan-Slavic thinkers had rejected. In the words of the veteran Bolshevik Nikolai Valentinov, writing in the 1950s:

> We were . . . attracted [to Marxism] by its European nature. Marxism came from Europe. It did not smell and taste of home-grown mould and provincialism, but was new, fresh, and exciting. Marxism held out the promise that we would not stay a semi-Asiatic country, but would become part of the West with its culture, institutions and attributes of a free political system. The West was our guiding light.[152]

Vladimir Ilyich Lenin, the leader of Russian Bolshevism until his death in January 1924, initially supported Trotsky's expansive programme in Asia and was also willing to consider incursions against the capitalist west in Europe.[153] In Asia, Lenin waged an undeclared war against the British Empire which would have been entirely familiar to Victorian intelligence officers of the 'Great Game'. Anti-British resistance movements were encouraged in Persia, and plans were drawn up for a central Asian army to invade India through Afghanistan, the scheme advocated before and after the Revolution by von Nieyermayer (see above). Lenin's invasion of Poland was partly designed to secure Russia's western frontiers but was also ideologically motivated: 'For the final victory of socialism', he once famously wrote, 'the efforts of one country . . . are insufficient; for that, the efforts of the proletarians of several advanced countries are required'.[154] Although he had previously dismissed the idea of a capitalist 'United States of Europe' as 'either impossible or reactionary . . . an attempt to retard America's more rapid development' and thus slow the emergence of globalised capital, Lenin nevertheless believed that revolution would ultimately sweep across Europe to create a solidly communist arena.[155]

The fact that the Revolution had occurred in peasant Russia rather than urban–industrial Germany presented an obvious problem for orthodox Marxists. Peasants, Marx had once claimed, were 'rural idiots' whose dependence on land (rather than wage labour) made them incapable of co-ordinated political action or genuine class consciousness. While the possibility of revolutionary overspill remained, this unfortunate fact could be overlooked. Once it became clear that revolution could not be exported, however, either to east or west, Lenin was forced to adjust his strategy and accept the Russian Revolution as an exceptional event which had defied Marxist predictions. Russia's economic and social conditions, its predominantly agricultural and peasant nature, coupled with its relative isolation from the 'mainstream' of European cultural and intellectual life, made it inherently vulnerable to counter-revolutionary forces, argued Lenin, both internal and external. These circumstances would have to change radically if Russia was ever to become a springboard for world revolution. Consolidating the revolution in Russia was, therefore, a more urgent task than exporting it to new lands: 'Our tactics ought to rest on the principle of how to ensure that the socialist revolution is best able to consolidate itself and survive in one country until such time as other countries join in.'[156]

In Lenin's view, the multinational character of the new Bolshevik state and the persistence of different nationalisms, particularly in those areas which had been briefly independent, was the source of internal counter-revolution. To reduce the build-up of nationalist resentment, Lenin insisted on a federal system of Soviet states, a Union of Soviet Socialist Republics (USSR), in which the distinctly Wilsonian ideal of 'self-determination' was explicitly enshrined. Lenin's belief in self-determination drew widespread criticism, notably from the anarchist Nikolai Bukharin who argued that this merely pampered to bourgeois, small-state nationalism which was implicitly anti-revolutionary. In his defence, Lenin argued that self-determination would always be necessary as an anti-imperial strategy beyond the Soviet Union and could not be rejected internally. According to Josef Stalin, then Lenin's ally, self-determination was 'a means in the struggle for socialism, subordinated to the principles of socialism'.[157]

Stalin was the least internationalist of the Bolshevik leaders. His struggle with Trotsky was, as Milan Hauner puts it, 'a geopolitical contest' based on a simple question: should the building of socialism be confined exclusively to the former territories of the Russian Empire, as Stalin believed, or should it be internationalised by exporting the revolution abroad, as Trotsky advocated?[158] Stalin's victory in this contest, complete by 1927, led to an intensification of the 'socialism in one country' strategy coupled with a rejection of Lenin's policy of tolerating national distinctiveness. Eventually, a centralised and repressive regime of staggering brutality was imposed which oversaw an exercise in wholesale economic and social engineering. For

Stalin, communism was not an ideology which could liberate or unite Europe, still less the wider world; it was simply an ideological mask to hide a totalitarian contempt for human life every bit as frightening as that of the Nazis.

Trapped by dogma and desperate to secure communism in the USSR, Stalin sought to reorganise the entire economic and social system not gradually, as Lenin had planned, but virtually overnight. More than a decade after the Revolution, the USSR had yet to reach the levels of industrial production achieved in 1913. Its industrial workforce numbered just 2.5 million and the vast bulk of the Soviet population, over 80 per cent, were still peasants. The USSR had to industrialise immediately, Stalin believed, if it was to produce its own proletarian class.[159]

The first task was to eradicate the entire class of larger farm-holders, the *kulaks*, who were evicted from their property and deported in huge numbers to Siberian labour camps. Some 200 000 were expelled from the Ukraine alone in the single year 1929–30. All farms, of whatever size, were then converted into 250 000 state collectives in an attempt to create a rural proletariat rather than a traditional peasantry. At least 120 million people were directly affected by these changes, and over 30 million peasant holdings were destroyed. By 1934, 90 per cent of the arable land of the USSR had been collectivised.

The resulting dislocation, coupled with ruthlessly enforced production quotas which forced starving labourers to surrender their meagre produce to the state, led to complete economic collapse. Around 18 million people fled the countryside for the cities from 1929 to 1935, despite strict regulations on population movement. Starvation and disease killed 5 million Ukrainians during the early 1930s (from a total rural population of just 20 million). Overall, perhaps 15 million people died in this savage experiment, a five-year war against the Russian peasantry which ranks as one of the greatest crimes in world history.[160]

The Communist Party under Stalin became an instrument of brutal centralisation, intolerant of all national variation. Each region in the Soviet federation was to become economically dependent on other regions, binding the union together through rigid central planning. The exploitation of Siberia's enormous mineral wealth, to be achieved with the assistance of slave labour, was a key component in Stalin's plans. The Party cultivated the idea of the heroic Communist man, overcoming with ease the environmental challenges of harsh continental interiors. An earlier generation of Russian Marxists had placed great emphasis on the formative role of the natural world, but Stalin forbade this, issuing a decree in 1938 making it illegal even to mention the influence of the physical environment on society.[161] The environment must be conquered and tamed by communism if the USSR was to build a new, self-sufficient realm across Eurasia. The appalling record of environmental abuse in the USSR and its satellite states is a direct legacy of this thinking.

Preoccupied with internal matters, Stalin and his acolytes gave little thought to external questions, actively discouraged geopolitical discussion and condemned this in other countries.[162] Socialist internationalism was barely mentioned, and the idea of a united Europe (as represented by Coudenhove-Kalergi and by Briand's 1930 memorandum; see above section on Europe reborn) was angrily denounced as a reactionary plot designed to undermine the USSR. Although their guiding ideology was quintessentially European, the attitude of Bolshevik leaders towards Europe was no more consistent than that of their Tsarist predecessors. Inspired by a system of thought which claimed universal significance, Russian Bolshevism proved incapable of producing a clear strategic viewpoint on any given region beyond the general assertion that, ultimately, the entire world must embrace communism. Stalin's desire to secure communism in the USSR produced an entirely introspective and authoritarian system of government.

Beyond the USSR, Russian exiles continued the pan-Slavic and Eurasian traditions of late nineteenth-century writers such as Nikolai Danielevskii (see Chapter 1). Amongst the most influential exiled Russian geopolitical theorists were Nikolai Alexsandrovic Berdiaev, Nikolai Trubetskoy, Petr Savitsky and Yurij Semyonov, each of whom settled in various western European capitals between the wars. Berdiaev, who lived in Paris, wrote an extraordinary book in 1924 on *The New Middle Ages* which extended Danielevskii's theme of Russia's unique spiritual status straddling east and west; a world unto itself, linked to Europe and Asia but separate from both. During the Middle Ages, Berdiaev argued, Russia had acted as the bridge between east and west and had thus facilitated the emergence of Europe out of Asia. The subsequent divergence of these two, naturally unified zones had been one of the more damaging consequences of the capitalist era and had also undermined Russia's unique position. The October Revolution, though dangerous in itself, provided an opportunity for Russia to lead the world on a journey of spiritual rejuvenation which would culminate with the reconnection of Europe and Asia, the beginning of a new 'Middle Ages' when the 'Russian idea' would triumph once again.[163] Trubetskoy sought to cement the fusion of Europe and Asia by proposing new hybrid languages. His *Europe and Mankind*, published in 1920, likewise saw the revolution as a necessary evil.[164] Savitsky, a geographer, also proposed 'practical' schemes to create a new Eurasian empire through a dense network of canals.[165] Semyonov, exiled in Germany and a close associate of Haushofer and von Niedermayer, insisted, like his German colleagues, on the continuity of Russia's geopolitical challenges and opportunities before and after the Revolution.[166]

The rape of Europa

Nazi preparations for war gathered momentum after 1933 in defiance of the post-1918 agreements. Supported by fascist Italy, Germany embarked on an ambitious rearmament programme which stimulated a temporary and probably unsustainable economic boom.[167] The outbreak of civil war in Spain between republican government forces and the fascist armies of General Franco in the mid-1930s was a harbinger of things to come, a clash between left and right in which the latter gradually prevailed with the active support of Italy and Germany whose military hardware was tested for the first time in combat. This tragic war, which claimed 600 000 lives and led to the emigration of 160 000 Spanish refugees, plunged Spain into four decades of authoritarian rule (Figure 3.13).[168]

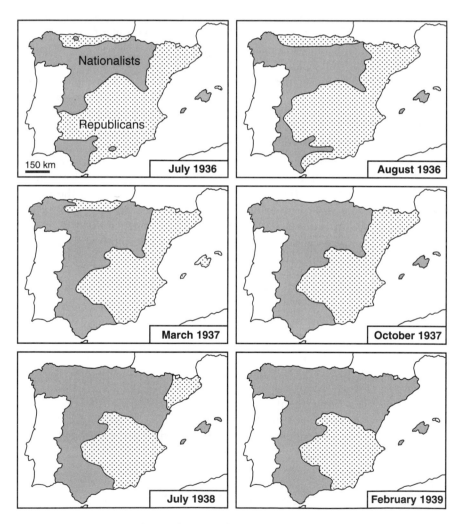

Figure 3.13 The prelude to disaster: the geography of the Spanish Civil War. Source: after Thomas (1977, 256, 402, 610, 732, 833, 885).

Germany's defiance was tolerated by the governments of Britain and France, who were desperate to avoid another war, and was even admired by some right-wing sympathisers who believed the Treaty of Versailles had been too harsh. The territorial dimension of German aggression manifested itself almost immediately. The disputed Saarland, which French scholars had tried so hard to claim for France in 1916–18, was ceded to Germany from League of Nations control in 1935 following a plebiscite. The Rhineland, which had been a demilitarised 'buffer zone' within Germany since 1919, was 'remilitarised'. In March 1938 an *Anschluss* was forced through by German and Austrian Nazis, uniting the two countries into a single *Reich*. Despite increasingly frantic negotiations, German troops marched into the mainly German-speaking Sudetenland of Czechoslovakia the following October on the flimsy pretext of protecting the local population from Czech 'persecution'. The remaining Czech provinces, Bohemia, Moravia and Slovakia, were occupied as German protectorates in March 1939, along with the mainly German-speaking Memel region of Lithuania. Czechoslovakia disappeared from the map, sacrificed in the hope that its passing would mark the end of German territorial demands.[169]

The signing of a non-aggression pact between Germany and the USSR on 23 August 1939 led to the effacement of Poland from the European map, invaded from the west by the *Wehrmacht* and from the east by the Red Army. Over the following months Soviet forces occupied the three Baltic states and also invaded Finland and Rumania, seizing the province of Moldavia from the latter (Figure 3.14). Britain and France declared war on Germany, which responded by moving its tanks and armies into defenceless Denmark, Norway, Luxembourg, Belgium and the Netherlands before launching a meticulously planned *Blitzkrieg* on France in the summer of 1940 which overwhelmed the massive French defences in just six weeks. Thus began a conflict which was to engulf not only Europe but the entire globe, providing, as Eric Hobsbawm has remarked, a painful 'lesson in world geography'.[170]

Hitler's war was launched in the name of the German people but, as we have previously noted, Nazi propaganda drew on a wider European rhetoric. This was by no means consistent, however, and the German invasion of the USSR in 1941 represented an important turning point in Nazi rhetoric on Europe. Before that date, in the era of the Hitler–Stalin pact, the idea of Europe was less frequently invoked in Nazi propaganda (though, as we will see, it was always central to the regime's racial policies). After that date, Hitler persistently claimed to be acting in Europe's name.

During the first era of the Hitler–Stalin pact, between the autumn of 1939 and the summer of 1941, both Nazi and Soviet officials were forced to perform an acrobatic somersault in their official attitudes to each other, previously characterised by mutual loathing. For this brief period, the pro-Russian ideas developed during the 1920s by Haushofer and von Nieyermayer enjoyed a minor revival in Germany. Von Ribbentrop, the German Foreign

Minister who had negotiated the non-aggression pact, even promoted
Haushofer's notion of a Eurasian bloc as the basis of a new Nazi foreign policy and used Haushofer's son, Albrecht, as an adviser.[171] The Nazi propaganda chief, Joseph Goebbels, also switched the spotlight of hate from
Soviet-style Bolshevism in the east to the evils of British and US capitalism

Figure 3.14 Eastern Europe, 1939–40. Source: after Bullock (1991, 722).

in the west. Hitler was presented at this juncture as the defender of a reborn Germany, reasserting its rightful place at the core of a reconstituted Europe whose unsustainable political geography, imposed by the liberal establishments of Britain, France and the USA in 1919, had been swept away.

By the spring of 1941 German domination of western and central Europe was complete, following the conquest of Yugoslavia and Greece (Figure 3.15).[172] The political geography of Nazi Europe was complex, even chaotic, the outcome of strategic military requirements and rigid Nazi dogma. 'Puppet' regimes were established in the areas not directly under German military occupation. France, the principal defeated power in the west, was divided into a German zone of occupation in the north, centred on Paris and flanking both the English Channel and much of the Atlantic coast, and a vassal 'substate' in the south, whose government was based in the sleepy spa town of Vichy, nestling on the banks of the river Allier north-east of Clermont Ferrand.[173] The Nazi objective was to keep France in a position of permanent impotence, incapable of ever launching a war of *revanche*.

The new 'father' of Vichy France was Phillippe Pétain, hero of the French defence of Verdun in World War 1, who was supported by a succession of supine prime ministers, notably Pierre Laval. The latter was especially keen to ingratiate himself with the Nazi leadership in the hope of establishing France as 'the favourite province of Germany' in the New Order.[174] Those of a charitable disposition claim collaboration was a genuine attempt to make the best of a tragic situation, to protect a defeated and vulnerable country from the worst excesses of German domination. But Pétain was reactionary, anti-democratic and authoritarian by choice rather than necessity. He interpreted France's defeat in 1940 as the outcome of a prolonged moral decay, deliberately engineered by socialists, communists and Jews. A moral revolution was required to rebuild France from its agrarian roots. Violently hostile to city life, which he equated with immorality and degeneracy, Pétain celebrated the simple innocence and timeless harmonies of peasant life, the basis of the true France. With the support of ultra-conservative elements in the Catholic Church, the Vichy regime sought to return France to its rural origins, to liberate the people from the corrupting influence of cosmopolitan Paris, now safely under German control. The republican slogans of *liberté, égalité, fraternité*, the legacy of the 1789 Revolution, were replaced by the Pétainist trilogy of *famille, pays, travaille*.[175]

In the enlarged German *Reich* a new and centralised administrative structure was devised (Figure 3.16) alongside different systems for the army, the SS, and the Nazi party (the latter organised according to a system of regions, or *Gaues*). The increased population of the new Germany was swelled further by the arrival of more than half a million ethnic Germans, mainly from the Baltic states, Bessarabia and Rumania. By June 1941, preparations were complete for an all-out attack on the Soviet Union, Hitler's war for *Lebensraum* in the east. Operation Barbarossa, as it was code-named, was the greatest military campaign ever waged on European soil: 200 German

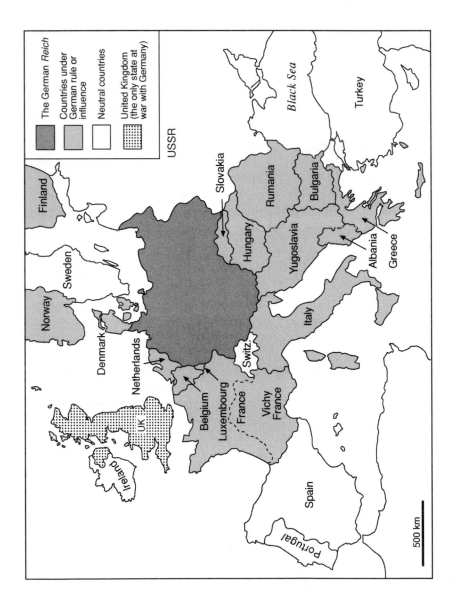

Figure 3.15 Europe, June 1941.

1. Schleswig–Holstein	13. Hesse–Nassau	25. Württemberg
2. Hanover	14. Thuringia	26. Bavaria
3. Mecklenburg	15. Lower Silesia	27. Tirol–Vorarlberg
4. Pomerania	16. Lorraine	28. Salzburg
5. Danzig–West Prussia	17. Westmark	29. Upper Danube
6. East Prussia	18. Hesse	30. Lower Danube
7. District of Bialystok	19. Sudetenland	31. Carinthia
8. Westphalia	20. Protectorate of Bohemia–Moravia	32. Styria
9. Saxony	21. Upper Silesia	33. Upper Carniola
10. Mark–Brandenburg	22. General Government	34. Lower Styria
11. Wartheland	23. Alsace	
12. Rhine Province	24. Baden	

Figure 3.16 Nazi Germany: the administrative geography. Source: after Freeman (1995, 59).

divisions attacked Soviet space along a 1500-kilometre front from the Baltic Sea in the north to the German–Slovakia border in the south (Figure 3.17). The initial objective was to force the Red Army behind a new front line running from Archangel on the White Sea coast in the north to Astrakhan at the mouth of the Volga on the Caspian Sea in the south. This would have given Germany over half the major towns and cities of 'European' Russia, including Moscow, which Hitler planned to destroy completely and replace with 'an artificial lake with central lighting'.[176]

As the German armies pushed eastward wholesale changes were made to the political geography of south central Europe. Yugoslavia was removed from the map and replaced by a complex geography of German and Italian occupied zones, with a fascist government installed in Croatia (Figure 3.18). Rumania, which had previously lost the province of Moldavia to the USSR, regained this region together with the province of Bessarabia to the north-west of the river Dniester (the pre-war frontier with the USSR) centred on

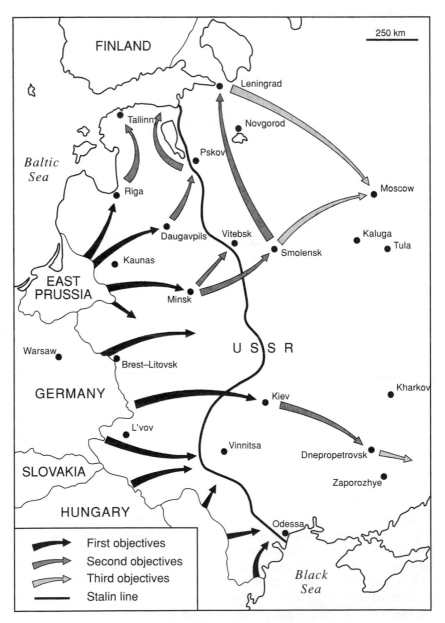

Figure 3.17 Operation Barbarossa. Source: after Bullock (1991, 803).

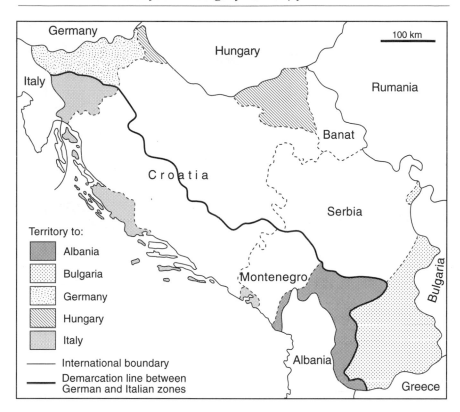

Figure 3.18 The end of Yugoslavia. Source: after Turnock (1989, 215).

the Black Sea port of Odessa. Transylvania, on the other hand, was ceded from Rumania to Hungary in 1940 and the region of Dobruja, south of the Danubian delta, passed to Bulgaria. The policies of all these 'countries' were, of course, entirely determined by their Nazi overlords. By November 1942, in the midst of the terrible war of annihilation in the east, Germany's territorial expansion was at its peak, its armies having swept across the southern reaches of the River Don, the ancient limit of Europe, south of Stalingrad (modern Volgograd) just as Hitler had predicted in 1937 (Figure 3.19; see also Figure 3.11).[177]

The invasion of the USSR in June 1941 unleashed an outpouring of Nazi vitriol against communism which was represented both as a Jewish conspiracy and a new form of Asiatic despotism threatening not only Germany but the whole of Europe. Suddenly Goebbels's propaganda was filled with visions of a reborn and united Europe, struggling under Germany's lead against a common eastern foe. The title of the Nazi periodical *Nation Europa* summed up the new geopolitical orthodoxy.[178] Nazi Europeanism was, to be sure, a profoundly immoral and empty geopolitical vision,

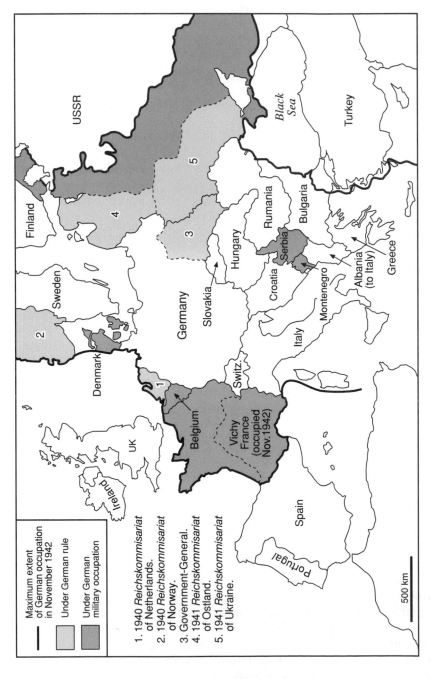

Figure 3.19 Europe, November 1942. Source: after Bullock (1991, 816–17).

founded on a traditional hatred of Russia as the home of Bolshevism and a sizeable proportion of Europe's Jews. Yet, tragically, this message was widely accepted far beyond the borders of the *Reich*. Appeals were launched for anti-Communists from other countries to unite with the *Wehrmacht* in its crusade. Special brigades were established within the invading armies from every European state, including an anti-Communist Russian unit led by a renegade Russian general who had defected. These brigades had little military significance but their propaganda value was considerable. As one Nazi propagandist put it as the tanks rolled into the USSR: 'Germany has rid the European continent of war'.[179] Hitler drew on the same European language, informing the German people in his address on 22 June 1941 that a new front had been opened in the east for the 'safeguarding of Europe and hence the salvation of us all'.[180] The Nazi declaration of war on the USA in late 1941 added an important anti-American and anti-capitalist subtheme to Hitler's new-found Europeanism.[181]

By 30 January 1942, Hitler's European message was as clear as it was mendacious: 'This is not a war we are fighting just on behalf of the German people. It is a struggle for the whole of Europe and the whole of civilized humanity'.[182] In the same month, the *Völkische Beobachter* informed its millions of readers that:

> Bolshevism will remain a latent threat to Europe so long as what is left of the Soviet armies has not been annihilated. . . . The soldiers of the Reich are not just defending the cause of their own homeland so much as protecting every European nation worthy of that name.[183]

Nazi interest in a united Europe culminated with the establishment on 5 April 1942 in Ribbentrop's Foreign Ministry of a 'Europe Committee' to oversee the creation of a federal, Nazi-dominated European Union which was due to begin operating the following autumn.[184] A month later Hitler informed an audience that 'the creation of a united Europe must remain the goal of our struggle'.[185]

Werner Daitz, the director of the *Gesellschaft für europäische Wirtschaftsplanung und Grossraumwirtschaft*, a Nazi-backed German planning agency, was one of the more passionately European Nazi theorists. In 1940 he wrote:

> The European community of peoples, the common *Lebensraum* of the white race, demands from each of its peoples the same discipline that the national community imposes on every one of its citizens. . . . [T]he peoples of Europe must again be Europeanized, so that they once more become citizens of their continent and, thereafter, of the world. Europe for the Europeans![186]

Two years later, Daitz was even more ecstatic, writing of the need for:

> the reorganization of European space . . . on the basis of natural laws,

so that it may again become a true enlarged *Lebensraum* of the European family of nations and therefore a genuine *Grossraum*. ... Once the new principle announced by the Führer, that race [*Volkstum*] and *Lebensraum* are supreme and inalienable values, is accepted by the whole European family of nations and finds expression in a new conception of law and a new legislation, states will be able to waive the exercise of many rights of sovereignty, as civil wars between the European family for the defence of *Lebensraum* will be largely unnecessary and consequently outlawed.[187]

Detailed plans were devised in several German ministries for a new European order which would transform the economic and social geography of the continent. One trivial but revealing example was Walter Christaller's work in Konrad Meyer's Planning and Soil Office on the *Generalplan Ost* (see above, p. 144). Christaller's 'Central Place Theory', familiar to generations of geography students and originally devised in 1933 as an explanatory model for the settlement pattern of southern Germany, became the model for a new system of towns and cities in the conquered territories of the east.[188] On the 'cultural' front, Himmler continued the plunder begun under Rosenberg by establishing the *Ahnenerbe* (Race Ancestry) 'research' unit of 'scholar-soldiers' within the SS which was to scour the galleries, museums and private art collections of the occupied territories stealing cultural relics and treasures which were deemed significant in the history of the German people.[189]

Despite the absurd language of a great European crusade, German interest in continental economic 'integration' after 1941 reflected the prosaic need to plunder Europe's resources to meet the massive demands of the attack on the USSR. Hitler made this abundantly clear in early 1942: 'This welding together of Europe has not been made possible by the efforts of a number of statesmen devoted to the cause of unification but by force of arms'.[190] Nazi economic 'integration', uncoordinated before 1941, was intensified after Albert Speer's promotion to succeed the recently deceased Fritz Todt as Armaments Minister in early 1942, but this was simply to meet short-term requirements. Indeed, Hitler was so contemptuous of wider considerations that he banned discussion of long-term economic planning in 1942.[191] His sole concern was the immediate needs of the army, and Speer was given *carte blanche* to use the whole of occupied Europe to ensure the German supply lines. Speer's remarkable 'success' in this mammoth task was built on the use of slave labour, co-ordinated by his sinister assistant Fritz Sauckel. Hundreds of thousands of non-Germans were forced to work in Germany itself or in German-owned factories beyond the *Reich*. In some parts of Germany over 15 per cent of the labour force were slave labourers from defeated countries.[192]

The misery inflicted on these unfortunates was nothing, however, compared with the fate of those groups which the Nazis decided had no place in the new European order, notably the Jews but also communists, gypsies, homosexuals and both the mentally and physically handicapped. As we have seen, the idea that Jews and these other groups should be removed not only from Germany but also from the whole of Europe had frequently been proposed during the 1930s in Nazi documents and propaganda. The passing of the Nuremburg Laws on 15 September 1935 depriving German Jews of virtually all their human rights was a clear indication of Nazi intentions for other areas of the continent which might fall under their control. Anti-semitism was, therefore, the most horrifyingly consistent aspect of Nazi 'Europeanism' and spread like a plague across the entire continent in the wake of Nazi expansion.

The exodus of German Jews began immediately Hitler came to power; 290 000 left before the outbreak of war, mostly to settle outside Europe.[193] This represented only the wealthiest strata of one of the more affluent European communities. Jews from the other countries which fell under the Nazi yoke after 1938 were less able to escape, hampered by the dislocations of war, their own poverty and the scandalous refusal of other countries, within and beyond Europe, to relax wartime immigration regulations, even for those with the necessary resources. Only 800 000 (fewer than one in seven of those eventually murdered) managed to escape from Europe between 1933 and 1945, the majority settling in the USSR (250 000), the USA (190 000), Palestine (120 000), Britain (65 000), Argentina (50 000), Brazil (25 000) and Paraguay (20 000). The overcrowded city of Shanghai, which required no visa for Jewish refugees, accepted more immigrants (25 000) than did Canada, Australia, New Zealand, South Africa and India combined.[194]

Nazi anti-semitism reached its peak following the German invasion of the Soviet Union when 'the final solution', as Herman Goering called it in May 1941, was developed and implemented. Europe's Jewish population was widely scattered across the continent, but the greatest number (44 per cent) still lived in Poland and the USSR, where Medieval Jewish migrants from the Byzantine Empire had established themselves and where Catherine the Great had created the 'Pale of Settlement' in the late eighteenth century as a special zone in which Jews were obliged to live. In several parts of Poland, Lithuania and the Ukraine, Jews represented over 15 per cent of the total population.[195] The staggering objective of the 'final solution' was the murder of Europe's entire Jewish population, beginning with those living in Eastern Europe. Rosenberg had announced in 1940, at the opening of a Nazi Institute for the Investigation of the Jewish Question, that 'Germany will regard the Jewish Question as solved only after the last Jew has left the Greater German *Lebensraum*'.[196] Just over a year later, on 9 October 1941, the dreadful implications of this statement were expressed in an even more sinister form by Hans Frank, the Nazi chief of the general government region of occupied Poland: 'As far as the Jews are concerned, I want to tell you quite frankly that they must be done away with one way or another'.[197]

This is what Daniel Goldhagen calls 'eliminationist anti-semitism'.[198] The extraordinary scale of Nazi plans is revealed by their estimates of the Jewish populations in each European country (including those which were either neutral or at war with Germany) as discussed at the Wannsee Conference in January 1942 where the plans for the 'final solution' were rehearsed (Table 3.2). Overall, some 14 million people were earmarked for execution.[199]

Table 3.2   Nazi estimates of Europe's Jewish population, January 1942.

Country or Administrative Region	Estimated Jewish Population
USSR	5 000 000
Ukraine	2 994 684
General Government (south-west Poland)	2 284 000
Hungary	742 000
Vichy France	700 000
White Russia	446 484
Eastern Territories (east Poland)	420 000
Bialystok District (north-central Poland)	400 000
Rumania	342 000
Great Britain	330 000
Occupied France	165 000
Netherlands	160 800
Germany	131 800
Slovakia	88 000
Bohemia and Moravia	74 200
Greece	69 600
Italy	58 000
'European' Turkey	55 500
Bulgaria	48 000
Austria	43 700
Belgium	43 000
Croatia	40 000
Lithuania	34 000
Switzerland	18 000
Serbia	10 000
Sweden	8 000
Spain	6 000
Denmark	5 600
Ireland	4 000
Latvia	3 500
Portugal	3 000
Finland	2 300
Norway	1 300

Source: Gilbert 1978, 14; Goldhagen 1996, 159

By the end of the war, six million Jews had been murdered, around 65 per cent of Europe's pre-war population, together with five million other victims whose lifestyles or beliefs were deemed unacceptable. Those responsible were initially the wandering members of four main *Einsatzgruppen*, established within Himmler's SS, to spearhead the slaughter in the wake of Operation Barbarossa with the assistance of local police battalions acting under their orders. Subsequently, as the scale of murder escalated, so a purpose-built infrastructure of mass killing was constructed, mainly in eastern Europe, comprising concentration camps, where victims were worked to death, and death camps, which were simply factories of slaughter. These

Figure 3.20 The infrastructure of murder: concentration and death camps. Source: after Bullock (1991, 834–5); M. Gilbert (1978, 16); Freeman (1995, 163).

were the nodes in an integrated European railway network on which millions were transported to their death from each corner of occupied Europe (Figure 3.20).

The Holocaust, as it is generally known, or the *Shoah*, as many Jews prefer, could not have been imagined, let alone implemented, without the Nazi domination of Europe.[200] But it is important to recognise the culpability of non-Germans too, either as active participants, facilitators or silent witnesses. Policemen across central and eastern Europe enthusiastically pursued and murdered Jews from 1939 to 1945, and soldiers and bureaucrats from every country in the region marshalled railway transports, often in the full knowledge that each wagon was filled to bursting with those condemned to die. In the west, the Vichy government in France passed heinous anti-semitic laws, partly to curry favour with Germany but also, bizarrely, to punish the group which were deemed to have contributed to the country's defeat by Germany in 1940. Far from being imposed by the Nazis, the virulent anti-semitism unleashed in France after 1940 was an indigenous phenomenon. The zeal with which the Vichy police rounded up and deported their Jewish compatriots may even have exceeded Nazi expectations. In all, at least 76 000 French Jews perished in the Holocaust.[201]

The proportion of Jews murdered varied considerably from country to country, depending on the length and viciousness of the Nazi occupation and the degree of local collaboration (Figure 3.21). Nearly 90 per cent of Polish Jews were murdered in the three administrative regions of the former state (the General Government, the Eastern Territories and the Bialystok District) and the vast majority of the pre-war Jewish populations were also killed in Germany, the Baltic States and Czechoslovakia. In Denmark, however, most of the small Jewish community (*c.* 8000 people) survived thanks to Danish rescuers who facilitated their escape to neutral Sweden. Death rates were also relatively low in Bulgaria and, most surprisingly, in Italy where the long years of fascist rule did not spawn the fanatical anti-semitism so evident elsewhere.[202]

Mapping the Holocaust is a difficult and dangerous task. While ransacking the 'data' on the most horrendous crime in history to concoct neatly ordered, two-dimensional cartographic images one can easily lose sight of the unspeakable barbarity of mass killing and the dreadful suffering of the victims. The maps thus produced are instantly recognisable from other contexts; they might be showing industrial output or unemployment rates, the stock-in-trade of conventional European geographies. Using these cartographic techniques for this very different purpose seems almost shocking, yet in a peculiar way, seeing the Holocaust geographically captures the dreadful bureaucratic banality of the evil on which it was based. Those who organised and perpetrated mass murder rarely stood out as monsters. They were 'ordinary' men and women, kind to their families and pets, model friends and neighbours. They were simply obeying orders, or so they claimed; they were mere cogs in a great administrative machine whose real

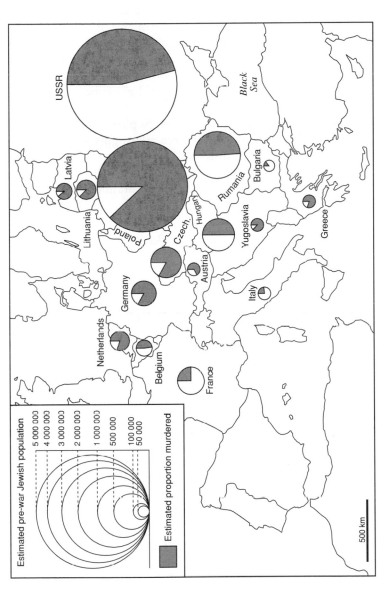

Figure 3.21 Mapping the Holocaust. Source: after Bullock (1991, 1088–9); Freeman (1995, 195); M. Gilbert (1978, 52). Note: this map should be treated with caution. It is based mainly on Alan Bullock's 'highest estimates' which are in turn derived from J. Noakes and G. Pridham, *Nazism, 1919–1945. Vol. III – Foreign Policy, War and Racial Extermination* (Exeter 1988), no. 918. The same source was used to produce Michael Freeman's map. These figures are a good guide in some cases but they can also mislead. Bullock's 'highest estimate' for France, for example, is 130 000, which would have produced a 43 per cent murder rate in relation to the 1940 Jewish population. This seems to be based on the official French government figure for the total number of people deported from France during the Vichy period, including Jews, gypsies and other groups. Marrus and Paxton (1981, 343–4) insist that around 76 000 Jews were deported, virtually all to their death. If this figure is used, the French murder rate in relation to the pre-war Jewish population falls to 25 per cent, as indicated here. Martin Gilbert's figures are different again. Although usually falling between Bullock's lowest and highest estimates, Gilbert claims 300 000 Hungarian Jews were murdered, which increases the murder rate in that country from 50 to 75 per cent. He also claims a somewhat higher total than Bullock's 'highest estimate' for Greece and a lower total than Bullock's 'lowest estimate' for Italy.

purpose was unknown or only dimly perceived. This is, of course, a terrible lie but it is one which we must confront on its own terms if we are to make sense of the geographical meaning of Europe. Claude Lanzmann, whose *Shoah* (1985) is the greatest cinematic attempt to engage with the Holocaust, has suggested that the massacre of Europe's Jew's needs to be understood in these apparently banal, bureaucratic terms for that is precisely how those who perpetrated this crime were able to rationalise their 'limited' contribution – as a minor administrative duty in a complex chain of command. *Shoah* does not rely on film or photographic evidence from the past but offers instead a nine-hour 'present-day geography' of the sites of former extermination camps, the homes and workplaces of surviving witnesses and other significant places where atrocities were planned or committed. A moving train is the film's leitmotif and there are lengthy scenes in which Lanzmann discusses with another historian the facts and figures of railway journeys and timetables, the 'problems' which the Nazis had to overcome in their attempt to maximise the use of a limited number of trains to transport Jews from Europe's ghettos to the death camps. These scenes encapsulate the terrifying managerial nature of the Holocaust. To those organising mass extermination, the task was simply to move 'cargo' from one place to another as efficiently as possible. This was a transport problem, a question of geography. It is for this reason that Lanzmann insists that his magnificent work is 'un film de topographe, de géographe'.[203]

The slaughter of Europe's Jews continued until the tide of Nazi domination had turned and the main nodes in the Holocaust network were overrun by the advancing Russian, US and British armies. Following their extraordinary resistance at Stalingrad (Volgograd) in late 1942, the Red Army gradually forced the German invaders back through 1943 and early 1944, while US and British forces landed in Sicily and fought their way northwards through Italy. In June 1944, US and British forces landed in Normandy in occupied France (Figure 3.22). With his empire collapsing around him, Hitler still ranted about the 'end of Europe': 'I'm the last hope of Europe!', he screamed, 'The new Europe will not be created through parliamentary decisions, nor through discussions and resolutions, but solely through force'.[204] By the following April, Nazi Germany had imploded from the west and the east. With the Red Army poised to take full control of the shattered city of Berlin, Hitler committed suicide in his underground bunker. A few days later, on 7 May 1945, Germany surrendered.

The cost of World War 2 in human life will never be precisely known. Millions expired leaving no record of their passing. A total of 46 million men, women and children, at least half of whom were civilians, is often cited. This translates to a death rate of more than 19 300 people per day from September 1939 to April 1945. Proportionally, Poland suffered more than any other country. Over 17 per cent of its 1939 population were killed. In absolute terms, however, Russia suffered most, with at least 21 million dead, 11 per cent of the spring 1939 population (Figure 3.23).[205] Indeed, of

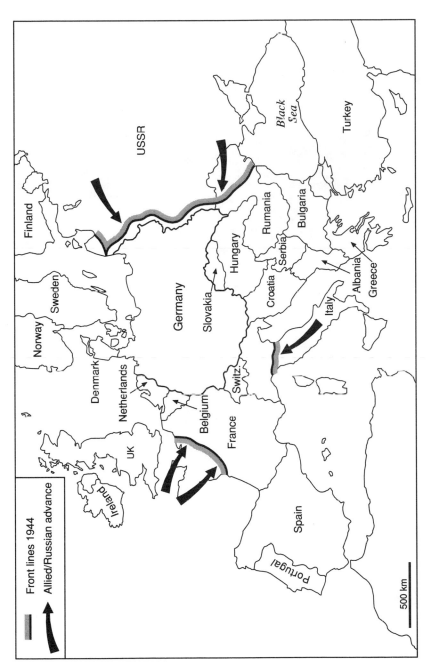

Figure 3.22 Europe, August 1944. Source: after Bullock (1991, 953).

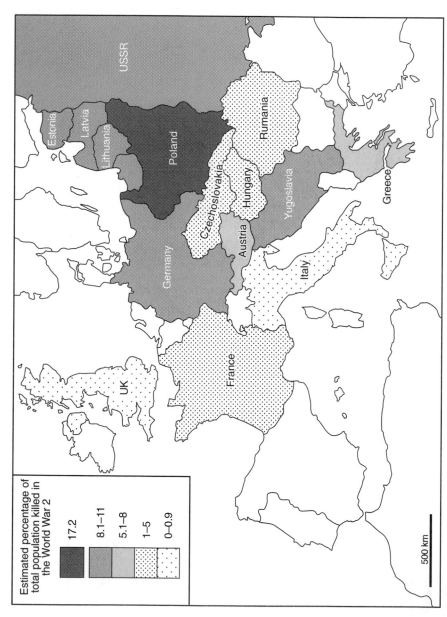

Figure 3.23 The lost generation: the impact of World War 2 on Europe's population. Source: after Bullock (1991, 1086). Note: this map shows the pre-war international borders.

the 60 million Europeans who died as a result of war and its associated horrors – genocide, disease, famine – between August 1914 and April 1945,
more than half were born either in imperial Russia or the USSR (and this
astonishing figure *excludes* the 15 million who died during Stalin's war
against the peasantry during the 1930s). Two-thirds of all Soviet towns and
cities lay in ruins in 1945 and a quarter of the country's capital assets had
been wiped out. This represented at least twice the material damage suffered
by Germany.[206]

Resistance and the European idea

The brutal implementation of Nazi geopolitics after 1939 prompted
geopolitical theorists in the western Allied countries to propose their own
theories of the post-war order. The ageing Mackinder revised his
'Heartland' thesis for the third time in 1943 in a more explicitly anti-
German way for Bowman's *Foreign Affairs*. Here, Mackinder looked
forward to a global peace-keeping alliance between the USSR and the
Atlantic powers.[207] The Dutch–American geopolitical commentator,
Nicholas Spykmann, proposed a further variation of Mackinder's arguments which likewise saw an Anglo–American–Soviet accord as vital for
the maintenance of order within Europe, a region he located in the geopolitically fragile 'rimland' encircling Mackinder's 'Heartland'.[208] Griffith
Taylor even offered a 'humanised' alternative to geopolitics, what he
termed 'geopacifics', which emphasised economic factors in the creation
of a stable geopolitical order. The war provided the opportunity for a
global stocktake, he argued, which should inform the new geopolitical
order and prevent war. Europe should be divided into four trading zones,
'crop-power blocks' delimited so that each had adequate supplies of tim-
ber, food, fuel and iron ore (Figure 3.24).[209]

The fascist appropriation of the European idea during World War 2 was
also directly challenged in Nazi-dominated Europe by the main resistance
movements. These were complex coalitions where conservative nationalists
rubbed shoulders with clerics and communists. Some resistance leaders,
such as Paul Bastid in France, were suspicious of any European rhetoric on
the grounds that this was now inextricably associated with Nazism.[210]
Others based their opposition to Hitler on the conviction that a united, federal Europe was the only way to guarantee liberal democracy. This was, to
be sure, partly a strategic view for a European resistance movement divided
into antithetical national groups would obviously be less effective against a
common Nazi enemy which seemed likely to dominate Europe for the foreseeable future. For this reason, the proclamations issued by the main resistance organisations, including the underground groups in Germany itself,
all spoke of the need for a united Europe to avoid war and to crush anti-
democratic ideologies.

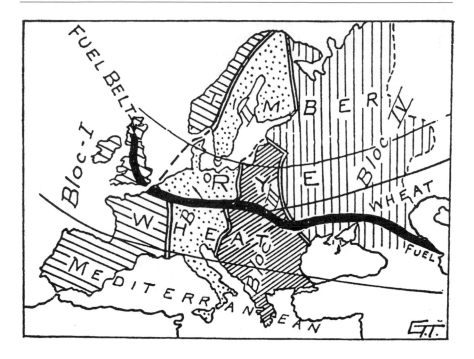

Figure 3.24 Geopacifics? Griffith Taylor's Europe, 1946.

A foundation document was the *Venetotene Manifesto* of July 1941 written by the Italian resistance leader, Altiero Spinelli. This led directly to the creation of the *Movimenta Federalista Europeo* (European Federalist Movement, or EFM) in August 1943, an alliance of different resistance and anti-fascist movements from across Europe. The EFM organised a conference in neutral Geneva in July 1944 from which emerged a second proclamation, again largely written by Spinelli, *The Manifesto of the European Resistance*, signed by nine main resistance movements. This called for the creation of a United States of Europe with a single parliament and army, to be established in a single bound the moment Nazi Germany and the other fascist regimes were defeated. Coudenhove-Kalergi, in exile in the USA, offered his support.[211]

Despite Britain's reputation as home to a relatively Euro-sceptical intellectual community, there was a remarkable wartime enthusiasm in the United Kingdom for a post-war European federation. Churchill, so often seen as the indomitable patriotic leader, had formally proposed an 'indissoluble' Anglo–French union on 16 June 1940, on the brink of France's collapse, with the ringing declaration that:

from now onwards France and Britain shall no longer be two nations but one Franco–British Union ... [with] joint organs of defence, foreign, financial and economic policies. Every citizen of France will enjoy immediately citizenship of Great Britain; every British subject will become a citizen of France.[212]

This was, to be sure, a desperate measure, with the Nazis poised on the brink of total European dominance, but it reflects the changed context in which Britain found itself in 1940. Isolated against an apparently over-whelmingly powerful enemy and home to many exiled European intellectuals, a substantial body of British opinion converted to a Euro–federalist viewpoint. The period 1939–41, after the outbreak of war and before the involvement of the USA, was probably the high point of British enthusiasm for European federation. One historian even suggests that

Despite the flow of material on the subject of European integration in the post-war years, it is doubtful whether such an impressive com-bination of quality and quantity has been produced in such a short period in any country since.[213]

The main forum was the Federal Union, established in the summer of 1939 by Charles Kimber, Patrick Ransome and Derek Rawnsley, and chaired initially by Ronald Mackay and then by Frances Josephy, Labour and Liberal politicians, respectively.[214] The movement attracted a remark-able alliance of politicians and civil servants (Lord Lothian, Sir Richard Acland, Richard Law, John Parker, Sir Drummond Shiels, Henry Usborne, Barbara Wootton and Konni Zilliacus), academics (Lancelot Hogben, Ivor Jennings, C.E.M. Joad) and journalists (H.N. Brailsford, Wickham Steed). Within a year it had over 12 000 members and 225 branches throughout the United Kingdom. In Oxford, the movement's spiritual home, a research institute was established at University College, headed by Sir William Beveridge and including the young Harold Wilson, which produced a steady flow of pamphlets. W.B. Curry's *The Case for Federal Union* (1939) sold over 100 000 copies in under six months.[215]

Others who were not members of the Union were also convinced. Clement Atlee insisted in November 1939 that 'Europe must federate or perish', and David Davies, the Welsh Liberal and founder of the 'New Commonwealth' movement, drew up his own detailed plans for a 'federated Europe'. In academia, the zoologist Julian Huxley proposed that all indus-trial capacity and resources be transferred from national to European con-trol, while G.D.H. Cole, the guru of British socialism, argued for a supranational European order divided into three great 'economic planning units'.[216] The Rumanian-born David Mitrany advocated *A Working Peace System* in 1943, 'a spreading web of international activities and agencies, in which and through which the interests and life of all nations would be gradually integrated'.[217]

Exiles from Nazi Europe in both Britain and the USA also contributed to the debate on the future of Europe.[218] One of the most intriguing organisations was the Committee of Fifteen, a small group of US-based European exiles, including Thomas Mann and Lewis Mumford, who met regularly from late 1938 to contemplate a new post-war Europe. Their 1940 pamphlet, *The City of Man: A Declaration on World Democracy*, eloquently demonstrates the triumph of utopian hope over brutal geopolitical realities.[219]

Summary

Despite the transformation of the political map in 1919 the underlying geopolitical assumptions which determined the new order in Europe were not radically altered by World War 1. As a result, alternative political geographies of Europe were developed through the 1920s and 1930s by commentators of liberal, fascist and communist views. Many, though by no means all, of these alternatives were predicated on the idea of European unity. The form this unity would take and the methods by which it was to be achieved were dramatically different for liberals, fascists and communists. Indeed, no consensus existed even within these three perspectives. Before 1914 and, as we shall see, after 1945, European unity was generally invoked for compatible reasons by its different advocates. Between the wars, however, this slogan was used for entirely different reasons by constituencies which were violently opposed to each other. The Nazi bid for European dominance in the late 1930s brought about the complete collapse of the European geopolitical order established in 1919. Many of those who fought against Nazism were committed to the idea of a united, democratic and liberal Europe. Their ideas were to develop into the post-1945 era.

Notes to Chapter 3

1 Bullock (1991, 1085) suggests *c.* 7.5 million *European* military deaths; M. Gilbert (1994, 541) argues for *c.* 8.3 million.
2 M. Gilbert (1994, x); Mosse (1990, 3–4).
3 G. Parker (1995, 208).
4 Winter (1988, 207).
5 Kitchen (1988, 22).
6 Hobsbawm (1994, 44).
7 Winter (1988, 195).
8 M. Gilbert (1994, 540).
9 Kitchen (1988, 22).
10 Bourke (1996); Diehl (1987); Whalen (1984).
11 Cannadine (1981); Gillis (1994); A. Gregory (1994); Heffernan (1995); Inglis (1993); Mosse (1990); Winter (1995).

12 Eksteins (1989); Fussell (1975); Hynes (1990).
13 F. Scott Fitzgerald, *Tender is the Night* (1986, 67–8).
14 On the failure of World War I to transform popular consciousness, see Pick (1993, 137–204) and Winter (1995, 228–9).
15 Henig (1995); Sharp (1991).
16 These last two objectives were in direct contradiction in Germany where the very real threat of revolution would obviously be exacerbated by the imposition of punitive peace terms. The brutal repression of the revolutionary left or 'Sparticists', led by Rosa Luxemburg and Karl Liebknecht, in 1918–19 led to bitter recriminations across the centre-left of German politics and polarised an already embittered population.
17 Simon (1939).
18 Foster (1989, 461–93).
19 Gravier (1919).
20 Temperley (1961, Vol. 4 207); P.J. Taylor (1993, 212); see also Lederer (1963); Wilkinson (1951).
21 Devine (1921).
22 Louis (1967, 135 and 160).
23 Wilson's 14th point recommended that a 'general association of nations must be formed under specific covenants for the purposes of affording mutual guarantees of political independence and territorial integrity to great and small states alike'; see Henig (1995, 76); Wilson (1919).
24 Quoted in Martin (1980, 87). For testaments from academics on the House Inquiry, see House and Seymour (1921).
25 On the efforts of Bowman and others to influence inter-war US policy, notably through the pages of *Foreign Affairs*, see N. Smith (1986).
26 Bullock (1991, 1087); Figes (1996, 773).
27 Demangeon and Febvre (1935); Schöttler (1995; 1997).
28 Kemp (1981); see de Lapradelle (1928) and Ancel (1938) for revealing inter-war French discussions of frontiers and borders.
29 Craig (1984); Fink (1989); Stoianovich (1976).
30 MacDougall (1978).
31 Trachtenberg (1980).
32 Adamthwaite (1980); Ross (1983).
33 Bowman (1921); Brunhes and Vallaux (1921); see also Fleure (1921); Newbigin (1920); Ogilvie (1922).
34 Hobsbawm (1994, 51).
35 Kitchen (1988, 28); Knox and Agnew (1994, 176–7).
36 Aldcroft (1997, 30–88).
37 Clout (1996, 25 and 47).
38 Keynes (1920); see also Aldcroft (1997, 1–29).
39 Churchill (1974, Vol. 3, 2341).
40 Aldcroft (1997, 89–120, especially 97).
41 Aldcroft (1997, 121–39).
42 Aldcroft (1997, 140–94).
43 O. Spengler (1918–22).
44 Carr (1981); Kitchen (1988).
45 See Stirk (1996, 18–50).
46 On European doubts about a League of Nations, see Agnelli and Cabiata (1919).
47 Pegg (1983); R. White (1989); Wiedemer (1993); see also O'Loughlin and van der Wusten (1990).
48 Quoted in Bugge (1993, 96).
49 Ageron (1975).
50 Coudenhove-Kalergi's commitment to democracy sometimes faltered. In 1923

he described 'our democratic era' as 'a miserable interlude between two great aristocratic epochs: the feudal aristocracy of the sword and the social aristocracy of the spirit', which represented his utopia; quoted in Stirk (1996, 26).

51 Quoted in Stirk (1996, 27).
52 Coudenhove-Kalergi (1923; 1943).
53 Bugge (1993, 101); Stirk (1996, 27).
54 Demangeon (1920).
55 Demangeon (1923; 1931; 1932); Claval (1994); G. Parker (1987).
56 Quoted in Greenwood (1992, 5). Churchill wrote the preface for Coudenhove-Kalergi's last memoir; see Coudenhove-Kalergi (1953, ix–x).
57 Herriot (1930, 152); quoted in Stirk (1996, 32); see also de Jouvenel (1930).
58 Navari (1991); Stirk (1996, 34–6).
59 Pegg (1983, 141); quoted partially in Stirk (1996, 35).
60 Pegg (1983, 140–8).
61 Krüger (1989); Stirk (1996, 37–8).
62 Morgan (1996); see also Boyce (1980); White (1991).
63 Boyce (1989, 66); Greenwood (1992, 5).
64 Boyce (1989, 79).
65 Levy (1996); Thomas (1977, 86–101).
66 Harvie (1991).
67 P. Hall (1996, 136–73).
68 Livingstone (1992, 271–90).
69 Fleure (1921, 6).
70 Bryce (1922); see also H.A.L. Fisher (1927); Harvie (1994, 42–3).
71 Lothian (1935).
72 Butler (1932; 1940); Cole (1947); Laski (1932); see also E.W. Gilbert (1939, 1952); Dickinson (1947).
73 Quoted in Boyce (1989, 79).
74 House of Commons Hansard (July 1934, col. 2339).
75 Curtis (1950); see also M.L. Smith (1989).
76 Fawcett (1933; 1940; 1941).
77 Mosse (1981, 4–6).
78 Penck (1925); Penck and Fischer (1925); see Herb (1997, 55–75 and 95–118).
79 Stirk (1996, 40–2); see also Delaisi (1929).
80 Stirk (1989, 127); Herb (1997, 58–60); see also Stirk (1994b).
81 Fahlbusch et al. (1989); Heske (1986; 1987); Korinman (1990); Kost (1988; 1989); Kristof (1960); Norton (1968); W.D. Smith (1986). Wartime considerations of Haushofer and German geopolitics include Dorpalen (1942); Gyorgy (1944); Strausz-Haupé (1942); Weigert (1941; 1942). On US paranoia about Haushofer, see Ó Tuathail (1996, 45–50 and 111–40).
82 Braun and Ziegfeld (1929–30); see also Herb (1997, 67–118).
83 Haushofer (1930–34); Raffestin et al. (1995, 119–56).
84 Quoted in Hauner (1992, 171).
85 Haushofer (1941).
86 Haushofer (1924); Hauner (1992, 172–4). Haushofer supported Indian independence, though he placed his faith not in Gandhi but in the like-minded, pro-German nationalist leader Subhas Chandra Bose, who fled to Germany in 1941.
87 Quoted in Hauner (1992, 176); see also Haushofer (1930; 1937; 1941).
88 Hauner (1992, 193–5).
89 Mackinder (1904, 436).
90 On von Niedermayer's role in plotting the downfall of British rule in India, see Hopkirk (1994) and Hauner (1992, 84–96 and 174).
91 Kost (1989); Sandner (1989).
92 Bowman (1942); N. Smith (1984).

93 Siegfried, 'Preface' to Ancel (1938, ix); Goblet (1936, 13). 'Spagyric' medicine is based on alchemy, magic and astrology; see G. Parker (1987).
94 de Martonne (1930).
95 Aulneau (1926, 8); see also Ancel (1936–45); Lemonon (1931); L'Héritier (1928); Ormsby (1935); for a post-war review, see Sinnhuber (1954). The idea of *Mitteleuropa* was also satirised in imaginative literature. Robert Musil's *The Man Without Qualities* (1930–32) is set in Kakania, a confused, contradictory realm in the middle of Europe, whose name was devised to imply 'Shitland'; see Kundera (1992, 443).
96 Ancel (1936); Haushofer (1927).
97 Ancel (1938, 188); also quoted in G. Parker (1987, 147).
98 Payne (1995, 3–79 and 441–95); Eatwell (1996, 3–29).
99 Nolte (1968, 385).
100 Nolte (1965).
101 Herf (1984).
102 Eley (1988).
103 Mosse (1980); Payne (1995, 147–211).
104 Griffin (1991, 44).
105 Stirk (1989).
106 Eatwell (1996, 33–88); Lyttelton (1973); Payne (1995, 80–128 and 212–44).
107 Cofrancesco (1985, 179).
108 Duhamel (1931).
109 Morgan (1990, 29); see also de Grazia (1992); Spackman (1996).
110 Confrancesco (1985, 182–3); Galbo (1996).
111 Atkinson (1996); Raffestin *et al.* (1995, 159–212).
112 Atkinson (1995).
113 Steinberg (1990).
114 Abraham (1981); Broszat (1987); Bullock (1991, 54–101, 150–192 and 241–84); James (1986); Turner (1985); see also Freeman (1995, 17–50) for a geographical reading.
115 Bullock (1991, 95).
116 Bullock (1991, 1082–3).
117 Bessel (1984).
118 O'Loughlin *et al.* (1994; 1995).
119 Bullock (1991, 1082–3); see also Childers (1983); Kater (1983).
120 Bessel (1987); Broszat (1981); Bullock (1962; 1991, 341–86 and 470–510); Corni (1990); Frei (1993); Kershaw (1987; 1993).
121 Kershaw (1993, 108–30).
122 Stirk (1989).
123 Bassin (1987b). Nazi racism caused unease amongst some academic geopolitical writers. Richard Henning wrote in the *Zeitschrift* in 1936: 'I protest against throwing geopolitics and race studies (*Rassenforschung*) into one pot'; quoted in Hauner (1992, 185).
124 Some Nazi theorists tried to reconcile the ideas of *Mitteleuropa* and pan-regionalism with Hitler's race policies. Carl Schmitt and Werner Daitz advanced the concept of the *Grossraüme*, a self-sufficient union of states held together by adherence to a common set of principles from which all foreign interference would be banned. Each *Grossraüme* would be dominated by a core state and people. The 'lesser' races had no rights to statehood and would exist as second-class minorities; see Schmitt (1941); Stirk (1989, 137–9).
125 Salewski (1985, 37).
126 Quoted in Stirk (1996, 66).
127 Mosse (1978) is a useful starting point.
128 Goldhagen (1996).

129 Cohn (1996). *The Protocols* was one of a handful of books which Nicholas II, the last Tsar, took into his Siberian imprisonment; see Figes (1996, 242).
130 Cohn (1996, 152).
131 Cohn (1996, 156 and 214–18).
132 Even fellow Nazis were bored by Rosenberg's account; Goebbels described it as 'an ideological belch'; see Bugge (1993, 125–7); Bullock (1991, 91); Roobol (1990).
133 Poliakov (1974).
134 Bridenthal *et al.* (1984); Burleigh (1994); Burleigh and Wipperman (1991); Müller-Hill (1988); Weindling (1989).
135 Salewski (1985, 37–8).
136 Ó Tuathail *et al.* (1998, 36–9).
137 M.L. Smith (1990).
138 Quoted in Stirk (1989, 141).
139 Bullock (1962, 368); Stokes (1980); Wright and Stafford (1988).
140 Trevor-Roper (1973, 40); see also Burleigh (1988).
141 Rosenberg (1927, 52) was thus able to claim that French overseas expansion was 'a purely militaristic imperialism which springs not from the internal necessity of the people, that is the need for living space, but rather from . . . a lust for plunder and adventure'; quoted in Stirk (1989, 140).
142 Rollins (1995); Shand (1984); on German industry more generally, see Gillingham (1985).
143 Strohmayer (1996).
144 Van der Vat (1997, 50–9).
145 Gill (1993, 251).
146 Hitler spent a mere three hours in occupied Paris but claimed his visit to Napoleon's tomb was the greatest moment of his life. 'Wasn't Paris beautiful?', he later remarked, 'But Berlin must be made far more beautiful . . . when we are finished in Berlin, Paris will only be a shadow'; quoted in Sereny (1995, 217).
147 Neumann (1996, 95–130).
148 Berki (1989).
149 Hauner (1992, 150–1).
150 Quoted in Hauner (1992, 703).
151 Trotsky (1971, 31); also quoted in Berki (1989, 51).
152 Quoted in Figes (1996, 141).
153 Lenin returned to Russia from his Zurich exile in the spring of 1917 with the assistance of the German government which gave him free passage across central Europe in a sealed railway carriage. Like some lethal virus, it was hoped he would spread revolution in Germany's faltering eastern enemy.
154 Quoted in Bullock (1991, 211).
155 These quotes come from a short pamphlet Lenin wrote in 1915 on 'the slogan for a United States of Europe'; see Berki (1989, 47).
156 Quoted in Figes (1996, 539).
157 Quoted in Bullock (1991, 108).
158 Hauner (1992, 207); Bullock (1991, 193–239).
159 Bullock (1991, 284–340).
160 Conquest (1986).
161 See Bassin (1992) on the environmentalism of Georgii Plekhanov, 'the Moses of the [Russian] Marxists' as Figes (1996, 146–7) calls him. Environmentalism was also central to Nikolai Ustryalov's quasi-fascist colonists who lived in communal self-denial in the wind-swept isolation of Harbin on the Manchurian plain, north-west of Vladivostok in modern-day China. Rejecting any form of internationalism, they championed the concept of 'National Bolshevism' which,

like National Socialism in Germany, celebrated heroic vitalism and the cult of youthful vigour born of a sturdy life in the wilderness; see Hauner (1992, 31–2); Stephan (1978).

162 Other Marxists also criticised western geopolitics. Karl Wittfogel, a leading intellectual in the German Communist Party (KPD), and Nikolai Bukharin, Trotsky's anarchist ally in Russia, both denounced German *Geopolitik* as bourgeois propaganda, inherently opposed to the USSR; see Wittfogel (1985); Ó Tuathail (1996, 143–51); Hauner (1992, 207).

163 Berdiaev (1924). *The Russian Idea* was the title of Berdiaev's other major work, published in 1946.

164 Hauner (1992, 60–1).

165 Hauner (1992, 199).

166 Hauner (1992, 193–4). In 1934 von Nieyermayer and Semyonov produced an important book on the geopolitical problems of the Soviet Union, *Sowjet Russland: Eine geopoliticische Problemstellung*, to which Haushofer contributed.

167 Overy (1982).

168 Thomas (1977). Fascist Spain remained neutral during the war, much to Hitler's annoyance, but it also spawned a geopolitical movement; see Vicens Vives (1940).

169 On Czechoslovakia's economic incorporation into Nazi Germany, see Prucha (1990).

170 Hobsbawm (1994, 24); see also Bullock (1991, 624–98); M. Gilbert (1989, 1–116).

171 Michalka (1985). Karl Haushofer's position was never secure, however, particularly following the 'disgrace' of his friend Rudolf Hess. Arrested by the Americans after the war, Haushofer and his wife committed suicide during the Nuremberg trials in 1946. Albrecht, a traditional conservative monarchist, became involved in the resistance movement and was executed on 23 April 1945 for his part in the plot to assassinate Hitler in July 1944; see M. Gilbert (1989, 672).

172 Stirk (1996, 51–82).

173 For a detailed map, see McMillan (1992, 134).

174 McMillan (1992, 140); Paxton (1972). Pro-German and fascist leaders in other countries were also busy collaborating in the hope of influencing the shape of the New Order, particularly Anton Mussert in the Netherlands, Vidkun Quisling in Norway and Leon Degrelle in Belgium; see Stirk (1996, 55–6).

175 Lebovics (1992, 43–50); Pollard (1998).

176 Bullock (1991, 810); see also M. Gilbert (1989, 195).

177 Gilbert (1989, 198–271).

178 Garton Ash (1993, 387).

179 Quoted in M.L. Smith (1990, 53).

180 Salewski (1985, 48).

181 Stirk (1990).

182 Quoted in Bugge (1993, 110).

183 Quoted in M.L. Smith (1990, 61).

184 Lipgens (1985, 122–32, 138–45 and 150–62).

185 Quoted in McElligott (1994, 128).

186 Quoted in Lipgens (1985, 81).

187 Quoted in Lipgens (1985, 110–11).

188 Rössler (1989). Christaller subsequently repudiated his Nazi connections and joined the German Communist Party (the KPD).

189 Nicholas (1995); Schama (1996, 75–81). On Italian attempts to influence Germany's apparent interest in European unity, notably through an international conference in November 1942 on 'The Idea of Europe', see Morgan (1990).

190 Quoted in Trevor-Roper (1973, 541).
191 Stirk (1996, 58).
192 Bullock (1991, 767–865); see also Gillingham (1991, 52–64) on the Nazi economy, and Sereny (1995) on Speer.
193 Barraclough (1984, 27–8).
194 Gilbert (1978, 27–8).
195 Pounds (1985, 89–91); Wanklyn (1940).
196 Quoted in Gilbert (1989, 130).
197 Quoted in Gilbert (1986, 213). Frank's diary suggests he had previously doubted the feasibility of mass extermination on logistical grounds. 'We cannot shoot 2,500,000 Jews', he wrote on 19 December 1939, 'neither can we poison them. We shall have to take steps, however, to extirpate them in some way – and this shall be done'; quoted in Gilbert (1989, 35).
198 Goldhagen (1996, 49–128).
199 Estonia, under German occupation since October 1941, was described at Wannsee in the following January as 'free of Jews', its 1000-strong community having already been murdered.
200 Goldhagen (1996).
201 Marrus and Paxton (1981, 343–4).
202 Steinberg (1990).
203 Quoted in Olin (1997, 1); see also Bauman (1985). Spatial considerations of the Holocaust include D. Clarke *et al.* (1996); Cole and Smith (1995); Freeman (1995) and Gilbert (1982). Work on the memory of the Holocaust has extended Lanzmann's contemporary spatial concerns; see Charlesworth (1994); J. Young (1993).
204 Quoted in McElligott (1994, 151).
205 See also M. Gilbert (1989, 745–7).
206 Hobsbawm (1994, 48).
207 Mackinder (1943); see Hauner (1992, 160).
208 Spykman (1944); see also Ó Tuathail (1996, 50–3).
209 G. Taylor (1946).
210 Stirk (1996, 60).
211 Lipgens (1985, 202–658); Stirk (1996, 67–8).
212 Quoted in Lipgens (1986, 5); see also Schlaim (1974).
213 Pinder (1986, 26).
214 Mackay (1940); see also Pinder (1986; 1989); Mayne, Pinder, Roberts (1990); Wilford (1980).
215 Curry (1939); see also Beveridge (1940); Brailsford (1939); Jennings (1940); Josephy (1944); Lothian (1935). For recent comments, see Lipgens (1986, 35–155) and Pinder (1989).
216 Lipgens (1986, 167–8, 183–5, 191–4 and 196–8); see also Ridley and Edwards (1944); Strachey (1940). For a recent commentary, see Weigall (1990).
217 Nelson and Stubb (1994, 77–97).
218 Lipgens (1986, 279–650).
219 Lipgens (1986, 790–3).

|4|

The European ideal? United in division, divided in unity

Introduction

The preceding chapters have considered the idea of Europe in different periods. Paradoxically, it is difficult to replicate this approach for the years after 1945 when concrete manifestations of European economic and political integration at last began to emerge in the form of legal treaties, intergovernmental institutions and pan-European organisations. Despite these achievements, the post-1945 rhetoric on European integration has been circumspect and muted compared with the sweeping ambitions of earlier times. This more prosaic approach reflected widespread revulsion at the excesses of fascist and Nazi geopolitics as well as the urgent need for a political system which could deliver the basic requirements of food, shelter and gainful employment. Economic necessities rarely featured in the European debate before World War 2; after 1945 they became crucially important. But the new pragmatism was also, and more importantly, a manifestation of the final collapse of Europe's global hegemony. Destroyed by a generation of warfare, the continent found itself controlled (like much of the rest of the world) by 'external', non-European superpowers, the USA and the USSR, whose interests and ambitions were global rather than peculiarly European (Figure 4.1). As we shall see, the Cold War division of Europe facilitated military and economic integration on either side of the ideological divide but it also placed obvious geopolitical limits on these twin and rival processes. The pan-European schemes of earlier generations, their endless debates about the meaning and geographical extent of Europe, had little relevance when these parameters were decided in advance. In a divided Europe that was no longer hegemonic the idea of Europe was inevitably constrained. This partly explains why there have been so few intellectual histories of the European idea since 1945.[1]

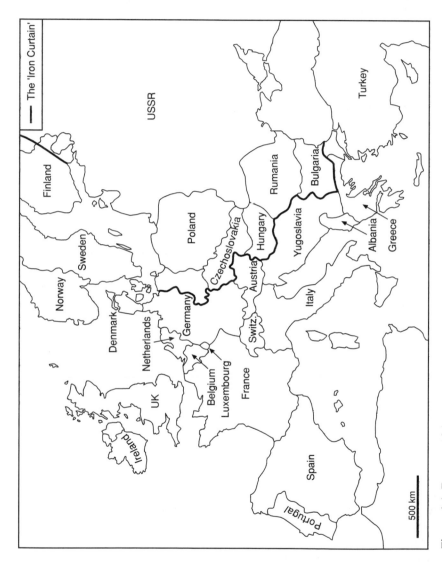

Figure 4.1 Europe, 1945.

This chapter examines the tensions and dilemmas associated with the processes of economic and political integration on either side of the divided continent from 1945 to the late 1980s. It also considers the challenge of creating a wider European agenda in the post-Communist era. To ease the reader's path through the jungle of acronyms, agencies and associations, a chronology of the major events in the process of European economic and political integration is laid out in Table 4.1.

Cold War geopolitics

The literature on the 'Cold War' – the global ideological stand-off between the US-dominated 'west' and the Soviet-dominated 'eastern bloc' – is enormous.[2] At the risk of oversimplifying, one can identify two rival interpretations. The earliest commentaries, mostly written by US authors, blamed the Cold War on the revival of Soviet expansionism after the Red Army's success against the Nazi war machine. According to this viewpoint, Stalin's 'socialism in one country' was abandoned out of necessity during the war and through conviction after 1945 when a more orthodox, Marxist–Leninist view of world revolution reasserted itself. Europe was now firmly in the sights of Soviet geopolitical theorists, and the USA together with its western European allies were simply responding to Soviet territorial ambition. The alternative view, promoted by critics of US foreign policy from the mid-1960s, argues that the Cold War was the geopolitical outcome of a post-war global capitalism dominated by US multinational corporations. From this perspective, the USA was the expansive and aggressive power, its foreign policy dictated by the world-wide demands of big business.

Most commentators operate somewhere between these two extremes. The concern here is not to attribute blame but to examine how a shared logic and a common language informed this ideological contest on both sides.[3] While the Cold War transformed older forms of nationalist geopolitics by creating a new, bipolar and transnational ideological confrontation, it was still based on a 'territorial imperative'.[4] This was equally discernible in both Soviet and US strategies and would have been instantly recognisable to geopolitical theorists from Machiavelli to Mackinder. Although operating at a global scale, the Cold War was still a struggle for space. In that simple but important sense it was a direct continuation of earlier forms of geopolitics.

The division of Europe into two zones in 1945 represented the first, crucial step in this new global contest. It owed little to negotiation and even less to the wishes of the European peoples. The dividing line was determined by the relative disposition of Soviet, US and British tanks at the time of the German surrender. The 'race' to seize as much central European territory as possible as the Nazi empire imploded demonstrates that both Soviet and

Table 4.1   A chronology of European integration from the Truman Doctrine
to the creation of the European Union

1947	March	Truman Doctrine announced
		Treaty of Dunkirk signed
	June	Marshall Plan announced
	October	General Agreement on Tariffs and Trade (GATT) established
1948	March	Treaty of Brussels signed
	April	Organisation for European Economic Co-operation (OEEC) established
	June	Eleven-month Berlin blockade begins
1949	January	Council for Mutual Economic Assistance (COMECON) established
	April	North Atlantic Treaty Organization (NATO) established
	May	Council of Europe established
		Federal Republic of Germany established
1950	May	Schuman Plan published
	October	Pleven Plan published
1951	April	European Coal and Steel Community (ECSC) established at the Treaty of Paris
1952	May	European Defence Community (EDC) treaty signed
	July	ECSC begins operation
1954	August	EDC rejected by France
	October	Western European Union (WEU) treaty signed
	December	United Kingdom becomes associated member of ECSC
1955	May	Warsaw Pact established
		WEU begins operation
	June	Messina Conference on 'further European integration'
	December	UK proposal for a 'Grand Design' on European cooperation
1957	February	United Kingdom proposes a European Free Trade Association (EFTA)
	March	Treaty of Rome signed
1958	January	European Economic Community (EEC) and Euratom begin operation
1960	January	EFTA established at the Stockholm Convention
	December	OEEC becomes Organisation for Economic Co-operation and Development (OECD)
1961	July–August	Berlin Wall erected
		Ireland, Denmark and United Kingdom apply for EEC membership
1963	January	De Gaulle vetoes British application to join EEC
		Franco–West German Treaty of Friendship signed
1965	July	France suspends involvement in the EEC
1966	January	'Luxembourg compromise' facilitates French return
	May–July	Common Agricultural Policy (CAP) agreed
	July	France withdraws from NATO
1967	May	Denmark, Ireland and United Kingdom re-apply for membership of the EEC
	July	ECSC, Euratom and EEC merge to become EC
	November	De Gaulle vetoes British application again
1968	July	Removal of internal customs duties within the EC
		Establishment of a common external EC tariff
		CAP begins to operate
1970	April	Treaty of Luxembourg signed
	August	Germano–Soviet Non Aggression Treaty signed
1973	January	United Kingdom, Denmark and Ireland join EC
	July	Conference on Security and Co-operation in Europe (CSCE)
1974	April	United Kindom demands renegotiation of terms of entry

	December	European Council (made up of the heads of national governments) agreed
		European Regional Development Fund established
1975	February	Lomé Convention signed
	March	United Kingdom renegotiation completed
	June	Greece applies for EC membership
	August	Helsinki Final Act of CSCE signed
1977	March	Portugal applies for EC membership
	July	Spain applies for EC membership
1979	March	European Monetary System (EMS) established
	June	First direct elections to the European Parliament
1981	January	Greece joins EC
1982	January	Common Fisheries Policy agreed
	February	Greenland votes to withdraw from EC
1983	June	Solemn Declaration of European Unity signed by EC heads of state
1984	January	EC and EFTA agree to establish common free-trade area
	February	European Parliament approves draft of European Union (EU)
	June	Revival of WEU
1985	February	Greenland withdraws from EC
	December	European Council adopts Single European Act (SEA)
1986	January	Spain and Portugal join EC
1987	July	SEA comes into effect
	August	Turkey applies for EC membership
1989	April	Delors report on European Monetary Union (EMU)
	June	European Council agrees first stage of EMU to begin in July 1990
	July	Austria applies for EC membership
	September–	Communist regimes collapse in Eastern Europe
	November	Berlin Wall destroyed
	December	Turkey's application to join EC rejected
1990	February	USSR endorses German unification
	October	German unification (former East Germany joins EU)
	December	Intergovernmental conferences on EMU and political union begin
1991	March	European Bank for Reconstruction and Development (EBRD) established
		Warsaw Pact dissolved
	June	COMECON disbanded
		Slovenia and Croatia declare themselves independent of Yugoslavia
	July	Sweden applies for EC membership
	December	Maastricht summit agrees treaty for political union and timetable for EMU
1992	March	Finland applies for EC membership
	April	War erupts in Bosnia
	June	Denmark referendum rejects Maastricht Treaty
	September	Britain and Italy leave Exchange Rate Mechanism (ERM)
		French referendum narrowly approves Maastricht Treaty
	December	'Opt-out clauses' agreed for Maastricht Treaty
1993	January	EC single internal market comes into existence
		Czech Republic and Slovakia become separate states in the 'Velvet divorce'
	May	Second Denmark referendum supports amended Maastricht Treaty
	July	United Kingdom ratifies Maastricht Treaty
	November	EU formally comes into existence

Anglo–American thinking was already determined by a Cold War territoriality. While both sides in the anti-Nazi alliance accepted that fascism could only be eradicated by an alliance of liberal democracy and communism, few believed that these two ideologies could coexist in peace and goodwill after the war. This sustains an argument, famously advanced by Ernst Nolte, that the period 1917–45 witnessed a continuous ideological struggle, focused mainly on Europe, between liberal democracy, communism and fascism. Although it was fought by rival nation-states, this was a titanic contest for the European soul which transcended national geopolitics. The destruction of fascism was bought at the cost of a Europe divided between the other two ideologies. The Cold War, so often interpreted as beginning in 1945, had much deeper historical roots.[5]

Yet, as Eric Hobsbawm has argued, there is no intrinsic reason why the anti-fascist alliance should have collapsed into such bitter competitiveness after 1945.[6] As we noted in the previous chapter, wartime proposals for a future European order were not based on traditional forms of political territoriality. Resistance leaders in Nazi-occupied Europe and their supporters in Britain believed in a post-war European federation, either a single overarching structure or several overlapping regional confederations. The details of these schemes varied but their objective was always the same: to retain the integrity of the wider European arena, both east and west. The failure of these ideas and the inability to imagine a peaceful *modus vivendi* between different systems after the war reflects the tenacity of territorial preoccupations on both sides of the ideological divide.

Wartime negotiations between US, British and Soviet leaders indicate that a bipolar world order was seen as the only possible outcome of a German defeat. While the August 1941 discussions between UK Prime Minister Winston Churchill and US President Franklin D. Roosevelt ended with a solemn declaration that neither the USA nor Britain would seek territorial aggrandisement from the war, their simultaneous insistence that the post-war order should be based on free trade presupposed the triumph of a US-style capitalism and was obviously unacceptable to the USSR.[7] Stalin was now convinced that if the USSR survived the German onslaught it would have to extend its influence over eastern Europe to protect itself from any future attack from the west. In December 1941, with the *Wehrmacht* poised at the gates of Moscow, Stalin informed Anthony Eden, the visiting British Foreign Secretary, that he would insist on major changes to the political geography of Eastern Europe once the German threat was removed.[8] The Allied plans for the invasion of Nazi Europe, discussed by Roosevelt, Stalin and Churchill at Tehran in November–December 1943, together with the various 1944 discussions on the future shape of Europe, intensified the territorial competitiveness within the anti-German alliance. The notorious 'percentages' agreement between Stalin and Churchill during the latter's visit to Moscow in October 1944 demonstrates 'east–west' territorial power games at their most cynical and manipulative. Churchill casually scribbled on a piece of paper the proportions of Rumania, Yugoslavia, Greece, Hungary and Bulgaria which he felt should fall under Soviet and Anglo–American influence after the defeat of Germany. According to this, the USSR would claim 90 per cent of Rumania, 75 per cent of Bulgaria and 10 per cent of Greece, the remainder of each country passing to the 'western alliance'. Yugoslavia and Hungary would be split 50:50. Both men knew that these speculations meant little at this uncertain stage in the war but the fact that valuable negotiating time was spent in such an exercise demonstrates the general acceptance that a post-Nazi Europe would be rigidly divided.[9]

From the autumn of 1941, a bipolar world with a divided Europe at its core seemed the inevitable consequence of an Allied victory.[10] The only question was where the dividing line would be drawn. The two major Allied conferences to decide the fate of Europe, at Yalta in the Crimea in February 1945 and at Potsdam six months later, sought to resolve this question. The Declaration on Liberated Europe, signed at Yalta, established the geopolitical changes which Stalin had described four years earlier in his discussions with Eden. Poland was re-established as Stalin insisted it should be, relocated to the west. The eastern part of pre-war Poland, beyond the 'Curzon line', reverted to the USSR while thousands of square kilometres of former German land, between the rivers Oder and Niesse, passed to the new Polish state (Figure 4.2). The USSR retained the Baltic states, which were seized in 1939, plus huge tracts of pre-war Finland, Czechoslovakia and Rumania. In addition, Rumania lost territory to Bulgaria.[11] The readjustment of eastern Europe's international borders gave the USSR a presence in the heart of central Europe and new land borders with most of the states in its sphere of influence (see Figure 4.1).

The crude territoriality of the Cold War reached its peak in the division of Germany and Austria. Plans for a divided Germany had been drawn up before the end of the war. The US Secretary of the Treasury, Henry Morganthau, wanted to render Germany permanently impotent by dismemberment and systematic de-industrialisation, a proposal more draconian than anything imagined after World War 1 and which mercifully came to nought.[12] The system eventually agreed at the Potsdam conference involved the division of both Germany and Austria into four unequal zones under the military occupation of US, Soviet, British and French forces who were charged with the huge task of rebuilding the war-shattered German infrastructure and ensuring the demilitarisation and de-Nazification of German society (Figure 4.3).[13] Stalin was persuaded that Berlin, firmly within the Soviet zone, should also be divided into four military zones, the three western sectors (occupied by US, British and French troops) to be supplied by restricted road, rail and air 'corridors' across Soviet eastern Germany (Figure 4.4).[14]

The four-fold division of Germany seriously affected the relative levels of economic development across the country. The strategies adopted by the occupying powers varied, despite attempts by the British and Americans to develop an integrated, 'bizonal' administration across both their sectors.

Established trade patterns and economic linkages between sites of production and traditional markets were seriously disrupted, and the administrative task was hampered by the movement of hundreds of thousands of refugees across the ideological divide from east to west. Some 30 000 migrants passed through Berlin each day in August 1945. There was also a vast movement of ethnic Germans across the new Polish–East German border. In all, 10 million Germans were homeless in 1945, a quarter of the European total.[15]

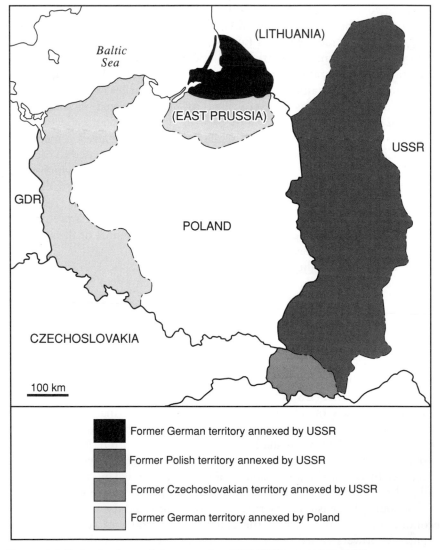

Figure 4.2 Poland relocated. Source: after C.H. Williams (1989, 210).

Figure 4.3 Germany divided. Source: after Blacksell (1977, 27).

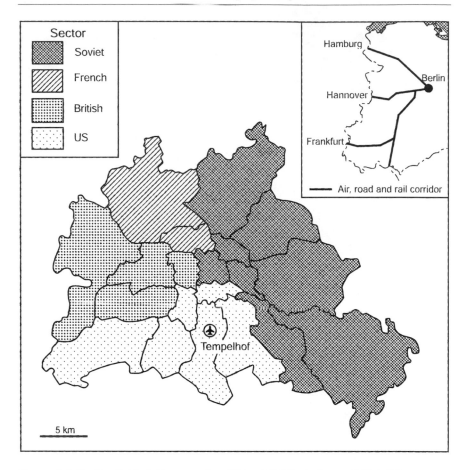

Figure 4.4 Berlin divided. Source: after Elkins (1988); A. Jones (1994, 37).

East is east and west is west

Within the geopolitical constraints imposed by the Cold War, the idea of an 'Atlantic Europe' became a new rallying cry for a growing number of western European and US politicians, quickly overwhelming the alternative vision, championed by several former resistance leaders, of a wider Europe made up of overlapping confederations. Churchill had been promoting an Anglo–American Atlantic alliance since the US entry into war, though he still believed in European integration.[16] Churchill's meeting with Roosevelt in 1941 (see above) led to the signing of the 'Atlantic Charter', a vague document which Churchill hoped would establish a 'special relationship' between the USA and the United Kingdom. This Charter, conceived at the darkest period of the war, echoed Churchill's forlorn attempt the previous year to establish an Anglo–French union against Germany (see Chapter 3), the obvious difference being the fundamentally unequal nature of any Anglo–American partnership.[17] The idea that an Atlantic 'Anglo-Saxon' alliance might coexist with a continental European federation was rendered far less likely by the military division of Europe in 1945.

In Britain, the incoming British Labour government, which toppled Churchill's administration in 1945, played a pivotal role in reinforcing the bipolar world and the division of Europe. Ernest Bevin, the new Foreign Secretary, cheerfully accepted the advice he was offered by the outgoing Prime Minister: 'The long term advantage to Britain and the Commonwealth', insisted Churchill, 'is to have our affairs so interwoven with those of the United States in external and strategic affairs, that . . . we stand and fall together'.[18] But, unlike Churchill, Bevin was suspicious of moves towards European political federation, though he accepted the need for greater economic integration. He was also especially hostile to the Soviet Union. Bevin sought to persuade the USA to adopt a more interventionist European and global policy which would, he calculated, protect the interests of an otherwise vulnerable and war-battered Britain and, only secondly, preserve the western half of Europe from the communist threat.[19] If the USA retreated into its traditional isolationism the British Empire would be left alone to confront a new and dangerous Soviet pan-region stretching, as Mackinder had predicted with such uncanny accuracy in 1919, from the heart of 'old Europe' across much of Asia. Even allied with the rest of western Europe, Britain would be no match for such an enormous land empire. As Peter Taylor puts it: 'A three world solution [the USA, the USSR and the British Empire] was Bevin's nightmare and given British scepticism for a one world this left a two world approach as Britain's only option'.[20] The US post-war *imperium* was, in this sense, an 'empire by invitation', with Britain acting as the principal facilitator.[21]

Churchill, now in opposition, continued his advocacy of the Atlantic option.[22] On 5 March 1946 he delivered one of his most memorable speeches in the otherwise unlikely setting of Westminster College in Fulton, Missouri, the very quintessence of small-town, mid-western America. This extraordinary geographical commentary, filled with the classical allusions of a public-school educated British aristocrat, was a deliberate attempt to strike fear into the hearts of the more isolationist elements in US political culture; to convince them that their real interests were best served by an expansive Atlanticism. Churchill's portrait of the new European order was littered with references to 'Communist fifth columns' operating at will and threatening the heartlands of 'Christian civilisation', most notably in the eastern Mediterranean where Greece, the well-spring of European culture, was especially vulnerable:

From Stettin in the Baltic to Trieste in the Adriatic, an iron curtain has descended across the continent. Behind that line lie all the capitals of the ancient states of Central and Eastern Europe. Warsaw, Berlin, Prague, Vienna, Budapest, Belgrade, Bucharest, and Sofia; all these famous cities and the populations around them, lie in the Soviet sphere and all are subject, in one form or another, not only to Soviet influence but to a very high and increasing measure of control from Moscow. Athens alone, with its immortal glories, is free to decide its future at an election under British, American and French observation.[23]

Soviet strategists were understandably dismayed by this clarion call for an Atlantic alliance to protect western Europe from communism. On 13 March 1946, in an article in *Pravda*, Stalin denounced Churchill's speech branding it (rather mysteriously) a sinister plot to replace Nazi race theory with an Anglo-Saxon equivalent in order to bolster Britain's crumbling imperial pretensions.[24] Many Americans were also unimpressed, the isolationists insisting that an America at peace with the world had no need to intervene in the affairs of Europe. A few were appalled by Churchill's anti-Soviet language. Henry Wallace, Roosevelt's Vice-President from 1940 to 1944 and Secretary of Commerce under Roosevelt's successor, Harry Truman, consistently emphasised the sacrifices which the Soviet Union had made in the struggle against Nazism. Paradoxically, the world had been made safe for democracy and US-style free trade through the efforts of the principal communist state. Refusing to co-operate with the USSR was morally indefensible: 'The tougher we get', claimed Wallace on 1 October 1946, 'the tougher the Russians will get'. US foreign policy should be expansive but 'vigorously dedicated to peace':

> To achieve lasting peace, we must study in detail how the Russian character was formed – by invasions of Tartars, Mongols, Germans, Poles, Swedes, and French; by the intervention of the British, French and Americans in Russian affairs from 1919 to 1921; by the geography of the huge Russian land mass situated strategically between Europe and Asia; and by the vitality derived from the rich Russian soil and the strenuous Russian climate.[25]

Wallace's enlightened idealism ran counter to the competitive, anti-communist Atlanticism which began to dominate US attitudes, particularly after Roosevelt's death in the spring of 1945. Cold War foreign policy gurus such as George F. Kennan were suddenly in the ascendant.[26] Kennan's vision of a 'global' America required a strong western Europe, a reliable second pillar in the Atlantic alliance against communism. This implied an economically and politically integrated zone, a western Europe modelled on the federal USA. Almost exactly a year after Churchill's 'iron curtain' speech, Truman revealed the volte-face in US thinking in an address to Congress which laid down the principles that were to govern US policy towards

Europe and the wider world for the next half century: the 'containment' of communism and economic expansionism, the twin elements in the so-called 'Truman Doctrine'. The context in which Truman made his announcement is crucial. British economic assistance to the beleaguered Greek government was due to end in just three weeks, on 31 March 1947, raising fears of a USSR-backed communist takeover. Rejecting the idea that the fledgling United Nations might intervene to prop up the Greek administration, Truman argued that only the USA had the power and wealth to prevent the fall of this strategically important country.[27] An immediate $400 million package of aid was recommended:

> I believe that it must be the policy of the United States to support free peoples who are resisting attempted subjugation by armed minorities or by outside pressures. If Greece should fall under the control of an armed minority, the effect on its neighbour, Turkey, would be immediate and serious. Confusion and disorder might well spread throughout the entire Middle East. Moreover, the disappearance of Greece as an independent state would have a profound effect upon those countries in Europe whose peoples are struggling against great difficulties to maintain their freedoms and their independence while they repair the damage of war.[28]

Truman's assessment of the USA's relative power was perfectly accurate. US military and economic muscle had been immeasurably strengthened by the war. The economy had grown by an unprecedented 10 per cent per year from 1940 to 1945, a rate never to be repeated. Industrial production had increased by a staggering 70 per cent and US factories were producing over half the world's manufactured goods by 1945–46. The USA also controlled 60 per cent of the world's oil, produced and consumed half the world's electricity and possessed half of all the telephones and radios on the planet. The USA was also the only country with a nuclear capacity, a fact revealed to a shocked world by the explosion of the first atomic bomb above the Japanese city of Hiroshima in August 1945.[29] But the other aspects of Truman's analysis were far more contentious, particularly the idea of states falling one after the other under the Soviet yoke, like so many collapsing dominoes. Fear of this dread spectre – part genuine, part cynical propaganda – played directly into the hands of the more aggressive proponents of US expansionism in the military and the corporate business sector.

Three months later, in a speech at Harvard University on 5 June 1947, George Marshall, Truman's Secretary of State, revealed the full scale of US global ambition and its new determination to protect 'vulnerable' states from communism.[30] Whereas Truman's 'modest' package of aid to Greece and Turkey had been concerned to preserve what were now being called 'western interests' in the Middle East and Europe, Marshall's proposal focused exclusively on the latter arena.[31] The Marshall Plan was the first attempt to establish an economically integrated Atlantic Europe. A massive

injection of capital was recommended over the period 1948 to 1951, mainly
in the form of free food and raw materials (the amount eventually made
available, *c.* $13.5 billion or 1.2 per cent of US gross national product, was
far below the $29 billion originally anticipated). This would allow the
European economies a breathing space to repair their war damage. The
explicit objectives were to increase industrial production, stimulate international trade, ensure financial stability and, significantly, to encourage
economic co-operation between European states.

Marshall's language was deliberately restricted to economic matters and
studiously apolitical. The war had produced a 'dislocation of the entire fab-
ric of the European economy', he argued, which now meant that Europe
imported twice as much by value from the USA as it was capable of exporting. This was in no one's interest, least of all the USA: 'Our policy is directed
not against any country or doctrine', he insisted, 'but against hunger,
poverty, desperation and chaos. Its purpose should be the revival of a working economy in the world so as to permit the emergence of political and
social conditions in which free institutions can exist'.[32] For this reason,
assistance would be offered to all European countries, on both sides of
Churchill's 'iron curtain'.

There was, to be sure, a genuine philanthropic motivation here.
European economic recovery was obviously crucial for the prosperity of the
wider world. The US experience of Roosevelt's 'New Deal' during the 1930s
suggested that ambitious, planned and co-ordinated modernisation was the
best solution to large-scale economic disruption. This would also contribute
to general political stability, an ideal embodied in the language of 'collective
security' which the new United Nations supposedly embodied. The
Marshall Plan was therefore presented as part of a politically neutral 'quest
for order', a manifestation of the US's non-ideological faith in 'the politics
of growth'.[33]

Few were convinced. The official Soviet response dismissed the plan as
flagrantly ideological, an attempt to bind together the war-weakened
economies of western Europe and ensure their collective dependence on US
capital. The offer was consequently rejected by the USSR in July 1947 and
the satellite states of eastern Europe were obliged to follow suit, to the dismay of the previously enthusiastic Czechoslovakian regime.[34] Andrei
Vyshinsky, the Soviet Deputy Foreign Minister, explained his country's attitude at the United Nations two months later:

> The so-called Truman Doctrine and the Marshall Plan are particularly
> glaring examples of the manner in which the principles of the United
> Nations are violated. ... [T]he United States government ... is
> attempting to impose its will on other independent states, while at the
> same time obviously using the economic resources distributed as relief
> to individual needy nations as an instrument of political pressure. ...
> This plan is an attempt to split Europe into two camps and, with the

help of the United Kingdom and France, to complete the formation of a bloc of several European countries hostile to the interests of the democratic countries of Eastern Europe and most particularly to the interests of the Soviet Union.[35]

US officials had predicted such a reaction and probably calculated their offer on this basis. If spread across the whole of Europe, both east and west, Marshall aid would have had relatively little impact. Bevin, anxious to secure the lion's share of US largesse, also wanted to exclude the USSR and was relieved by their rejection of the offer.[36] In the end, a quarter of Marshall aid went to the United Kingdom, one fifth to France and just over 10 per cent to both western Germany and Italy.[37] Opposition to the Marshall Plan was not limited to the USSR. Many western European socialists, particularly those in France and Italy who had advocated an independent federal Europe, echoed the Soviet interpretation.[38]

Whether the Marshall aid initiated European economic recovery is a moot point. European economic performance had certainly begun to stagnate by 1947, industrial production actually falling in 1946–7 when it was still only at 75 per cent of 1938 levels.[39] Without an infusion of external capital, production might well have slowed further. In the event, growth rates through 1948–51 were impressive, industrial production increasing by one-third, inter-regional trade by 70 per cent, exports by 68 per cent and imports by 20 per cent. Yet western Europe might have recovered under its own steam and without the long-term inflationary pressures which are sometimes attributed to the 'artificial' injection of Marshall funds.[40]

The main geopolitical implication of the Marshall Plan was its emphasis on *co-ordinated* European economic recovery. This objective was initially frustrated by reluctant European governments who were dedicated, in Peter Stirk's phrase, to 'extracting as much aid as possible and minimizing the extent of co-operation demanded by the United States'.[41] Nevertheless, the 'Americanisation' of western Europe's economies was advanced by the creation, on 16 April 1948, of the Organisation for European Economic Co-operation (OEEC), an international economic planning agency which linked together Austria, Belgium, Denmark, France, Greece, Iceland, Eire, Italy, Luxembourg, the Netherlands, Norway, Portugal, Sweden, Switzerland, Turkey and the United Kingdom, together with representatives of the Commanders-in-Chief of the US, British and French zones in western Germany. The activities of the OEEC expanded geographically and conceptually over subsequent years but its effectiveness was always limited by the domestic priorities of the constituent nations. The newly established Federal Republic of Germany joined in 1949 followed by the USA and Canada in 1950, reinforcing the organisation's Atlantic agenda.[42] The OEEC was inextricably associated with US economic and geopolitical values, its various subcommittees openly promoting US-style capitalism. The European Productivity Agency, for example, established within the OEEC in 1953,

sought to encourage the adoption of Fordist industrial techniques throughout western Europe. By 1960, the OEEC had transcended its original
European role and was renamed the Organisation for Economic Co-operation and Development (OECD), one of several post-war international economic agencies, including the International Monetary Fund (IMF) and the
World Bank, which championed US neo-liberal economic theory.[43]

The second component of an Atlantic Europe was the military alliance
between the western European states and the USA. Once again, the idea
arose from British concerns, which were shared by France, Belgium, the
Netherlands and Luxembourg, about the ability of the United Nations to
preserve the region's 'collective security'. Uncertainty about the USA's long-
term commitment to western Europe, notwithstanding the announcement
of the Truman Doctrine and the Marshall Plan, provoked these five governments to sign the Treaty of Brussels in March 1948, establishing a military
alliance against a future attack from the east. Unnerved by these and earlier
developments, the USSR sealed off the road and rail links to western Berlin
from late June 1948 to late May 1949, obliging the western armies to supply the two million inhabitants in their sectors of the divided city by round-
the-clock airlift. Partly as a result, the USA and Canada joined the Treaty of
Brussels alliance in October 1948. Over the next six months the agenda
shifted from a rather traditional European alliance of nation-states towards
a wider military 'federation' linking both sides of the Atlantic. The signing
of the North Atlantic Treaty in Washington on 4 April 1949 by the five
Treaty of Brussels states plus the USA, Canada, Denmark, Iceland, Italy,
Norway and Portugal created a new military pact, the North Atlantic
Treaty Organization (NATO). According to article five of this treaty:

> The Parties agree that an armed attack against one or more of them in
> Europe or North America shall be considered an attack against all,
> and consequently agree that, if such an armed attack occurs, each of
> them will assist the Party or Parties so attacked by taking forthwith,
> individually and in concert with other Parties, such action as it deems
> necessary, including the use of armed force, to restore and maintain
> the security of the North Atlantic area.[44]

Like the OEEC, NATO membership increased through the 1950s to
embrace Greece, Turkey and the new Federal Republic of Germany. Despite
this geographical expansion, NATO's agenda was dominated by the USA's
global strategic considerations, particularly the new Cold War confrontations
in the 'Third World' which began with the Korean war in the mid-1950s.[45]

The conviction that western Europe was set on an inevitable course
towards a US-style federation was sustained by the so-called 'functionalist'
orthodoxy in US geopolitical and economic writing during the 1950s and
1960s. According to many US economists, a western Europe in which
national rivalries persisted would never be able to deliver US levels of affluence, as would become increasingly obvious to the 'free' peoples of the

continent. Their only 'rational' option was to replicate the process of federal state-building which had created the USA. Once the youthful offspring of old Europe, the USA was now the mature exemplar for a reborn western Europe. The evolution towards federalism was frequently described in stadial terms, akin to those invoked by W.W. Rostow in his widely debated text *The Stages of Economic Growth* (1960).[46] Rostow's generalised schema produced comparable four-stage predictive models for western Europe, each stage supposedly delivering higher levels of production and consumption and thus reinforcing US-style capitalism. First, a western European free-trade association was predicted (emerging perhaps from the OEEC) which would allow tariffs and trade barriers to be lowered or removed on some or all products traded between the members states. Each member would be free, however, to establish bilateral economic and other accords with states beyond the association. This would logically evolve into a western European customs union in which all restrictions between partner states on the movement of labour and capital would be removed to create a single, common market. Member states would still be free to buy and sell outside the customs union but the trade in some products or services would be controlled by common economic policies. This would beget a western European economic union based on agreed economic policies limiting the 'national' freedom of states to trade independently of the union. This implied common policies, notably on transport and exchange rates, leading inevitably to a single currency and a central bank. The final stage would be a western European political union, a United States of Europe, in which all political rights and sovereignties would be pooled in an elected parliament with a single civil service and executive. Neo-functionalists, associated with the political scientist Ernst B. Haas at the University of California at Berkeley, developed a more sophisticated reading of this process. Haas argued that, once established, international institutions tend to assume their own internal dynamic which propels them towards greater integration even if that was not their initial objective. The concept of 'spillover', whereby co-operation in one area produces co-operation in another, was central to neo-functionalist arguments. European integration was still seen as inevitable, but this derived less from conscious decision-making than from the logic and quotidian demands of administering complex democracies.[47] Functionalist and neo-functionalist accounts drew on a managerial, technocratic and 'problem-orientated' rhetoric which appeared politically neutral but was self-evidently ideological. The subtitle of Rostow's book, 'A Non-Communist Manifesto', betrays the ideological values which informed functionalist theory.

In the communist world, attempts were made to replicate developments in western Europe, to create a communist 'Eurasia' to confront a capitalist 'Atlantic Europe'. The 'eastern' response to the OEEC was the Council for Mutual Economic Assistance (COMECON), established in January 1949 to foster trade and economic integration within a centrally planned, socialist system between the USSR, Bulgaria, Hungary, Poland, Rumania and Czechoslovakia (with Albania and the German Democratic Republic joining later).[48] COMECON was largely a propaganda exercise, an attempt to demonstrate that Soviet-style communism was capable of producing shared, democratic co-ordination between sovereign states. In fact, the economies of eastern Europe were controlled by Communist Party machines directed by Moscow. There was no political or economic integration in the sense of pooled sovereignty or tariff elimination, merely the imposition of a single economic system based on collectivised agriculture and the massive expansion of heavy industry. Little attention was paid to the social or environmental costs of this strategy, or to pre-existing economic conditions and market orientations. Indeed, the objective was to weaken previous ties and ensure that each eastern European state was totally dependent on the USSR. The biggest 'challenge' was the German Democratic Republic, a new state which had been dependent for its food, energy and industrial produce on the western half of the old Germany. The construction of a massively fortified and impenetrable border, including a notorious wall which snaked across the city of Berlin from 1961, demonstrated Soviet resolve.[49] Soviet strategy in eastern Europe was simply the export of 'domestic' policies used to control the different republics within the USSR, what Graham Smith calls 'federal colonialism'. This was characterised by the total domination of each national Communist Party and the redistribution of economic resources to ensure loyalty and dependence.[50] COMECON had virtually no role in this: its Assembly did not convene at all between November 1950 and March 1954, and the organisation had no formal charter until 1960.[51]

The USSR's dominance was underlined in May 1955 by the signing of the Warsaw Pact, the Communist response to NATO. This was a reflection of the rapid expansion of Soviet nuclear capacity (the USSR had developed its own intercontinental ballistic missiles by 1957) and the establishment in 1954 of an independent Federal Republic of Germany in the west which immediately joined the NATO alliance. Attempts to challenge the Soviet behemoth in eastern Europe met with immediate and violent response. The Hungarian experiment of a mixed economy and greater personal freedom was brutally suppressed by unilateral Soviet action in 1956 and a similar project, conceived 12 years later by the Czech leader, Alexander Dubcek during the so-called 'Prague Spring', was crushed by tanks from across the Warsaw Pact.

The military alliances on either side of the divided Europe played their part in the massive escalation in both conventional and nuclear weaponry. Expenditure on 'defence' increased by around 800 per cent in both the USA and the USSR from 1948 to 1969.[52] By 1975, the USA had built over 1000 intercontinental ballistic nuclear missiles, and the Soviet arsenal included over 1500 of the same, plus a further 600 intermediate-range bombs.[53] This was sufficient to destroy all life on the planet many times over, and Europe lay at the heart of this global stockpile, 'a walnut in the nuclear nutcracker'.[54] By the mid-1980s there were approximately 10 000 nuclear warheads on European soil, about half the world's total.[55]

European ideals and realities

The failure of a pan-European perspective in the immediate post-war years, when Europe's division was sealed by Cold War warriors in Washington and Moscow, also reflected the intellectual confusions and naïvety of the federalist lobby. There was no shortage of pressure groups advocating a wider European federation, many arising from wartime resistance movements, but these developed distinctive, often mutually exclusive, strategies.[56] The European Union of Federalists (EUF), established in the autumn of 1946, valiantly attempted to hold together 80 federalist associations from 13 countries in a single umbrella organisation, but ideological tensions undermined the prospects of a common programme. Left-wing groups, such as the Italian *Movimento Federalista Europea*, sat uncomfortably alongside liberal associations such as the British Federal Union (see Chapter 2) and the Swiss *Europa-Union*. The French organisation, *La Fédération*, was associated with the far right and included several former members of the fascist *Action Française*.[57] Federalist organisations operating outside the EUF were equally varied and ineffective. The *Mouvement Socialiste pour les États-Unis d'Europe* was a left-wing grouping which, despite its French title, was established in London in February 1947 at the instigation of the British Independent Labour Party, an organisation which distinguished itself from the main Labour Party by its radical internationalism.[58] The economic argument for federalism was advanced by the European League for Economic Co-operation, which included Paul van Zeeland from Belgium and William Beveridge and Harold Butler from the United Kingdom.[59] A more conservative, Christian Democratic message was advanced by the tireless Coudenhove-Kalergi who revived his Pan-Europa movement in the shape of the European Parliamentary Union.[60] This merged in 1951 with the only organisation which had any significant influence, the European Movement. This was created amidst much fanfare in The Hague in May 1948 to promote the idea of a European Assembly.[61] The dominant figure at the opening conference was Winston Churchill who, together with his son-in-law Duncan Sandys, helped raise the profile of the European Movement. Churchill, as we have seen, was committed to an Atlantic Anglo–American alliance but also believed in a pan-European federation on the continent. He had advocated a Council of Europe in March 1943 and sketched out his vision in greater detail in a speech in Zurich on 19 September 1946. This called openly for a United States of Europe to solve 'the tragedy of Europe ... that series of nationalist quarrels, originated by the Teutonic nations'. This should be anchored around a new Franco–German alliance, he argued, which was in the best interests of Europe, the British Empire and, by implication, the Atlantic alliance with the USA. Britain, the USA and even the USSR would stand aside from this development, he claimed, as 'friends and sponsors of the new Europe', rather than as members.[62]

The European Movement was successful in its immediate objective in that a Statute of the Council of Europe was signed within a year of the Hague conference, on 5 May 1949, by the governments of Belgium, Denmark, France, Eire, Italy, Luxembourg, the Netherlands, Norway, Sweden and the United Kingdom.[63] This document looked forward to 'a greater unity between its Members for the purpose of safeguarding and realising the ideals and principles which are their common heritage and facilitating their economic and social progress'.[64] Precisely how this was to be realised was never made clear. While most governments were willing to accept the Council as an intragovernmental 'think tank' which might generate policies on matters of general European importance, notably on human rights and latterly on environmental questions, the original terms of reference were extremely vague and specifically excluded any discussion of defence. The strictly limited role of the Council of Europe reflected the Atlanticist, Cold War agenda of those, Churchill included, who were largely responsible for its creation as well as the reluctance of most national governments to cede power to a European institution.[65]

Those who hoped for a united and independent Europe which might ultimately transcend the ideological division established in 1945 were understandably disappointed by the Council of Europe. Their room for manoeuvre was limited, however, as the levers of political power were closely controlled by national governments acting in the name of traditional nationalist interests. Paradoxically, a federalist programme required the support of national governments. The relaunching of a wider European project after 1950 was made possible by the emergence of a perceived convergence between national and European interests, first in France and subsequently in other western European states. France's pivotal role in this process reflected peculiarities in French national culture that can be traced back to the Revolution of 1789 and its Napoleonic aftermath. As we saw in Chapter 1, this era witnessed an ambitious attempt to create a new Europe in the image of revolutionary France. Although this failed, a conviction that the idea of France was somehow the quintessence of the idea of Europe lingered in French political and intellectual debate and has exerted a discernible influence on the country's foreign and colonial policies down to the present.[66] The rhetoric of French history and culture made it possible to equate national and European interests in a way which was simply impossible in the other European countries in 1950. British politics were still dominated by extra-European concerns connected with the empire and maritime trade. These interests determined the nation's cultural assumptions and shaped its educational system. Although the economic importance of the Empire/Commonwealth was in rapid decline, the emotional connections remained strong across the political spectrum. In the words of the National

Executive of the Labour Party in 1951: 'we in Britain are closer to our kins-
men in Australia and New Zealand on the far side of the world, than we are
to Europe'.[67] West Germany was also in no position to pioneer a European
project. Its last attempt do this had been inspired by Nazism and had ended
with the division not only of Germany itself but of the entire continent.
Italy's immediate past was scarcely less disastrous.

The conviction that a reborn Europe could be equated with a reinvigorated France was perfectly encapsulated in the figure of Jean Monnet, chief
of the National Planning Authority. Unlike the leaders of the federalist
organisations mentioned above, Monnet recognised that European integration would only be possible if national governments were persuaded of its
benefits. A European agenda had to coincide, or appear to coincide, with a
nationalist agenda. Once established in France, this equation might diffuse
across the whole of western Europe. Monnet convinced the French Foreign
Minister, Robert Schuman, who announced on 9 May 1950 that a united
Europe was the only guarantee of peace and was therefore France's official
policy, an apparently dramatic intervention which in fact merely restated
Aristide Briand's proclamation exactly 30 years earlier (see Chapter 3).[68]

Schuman's proposal argued for an economic alliance between France and
Germany which would so link the two countries as to make conflict between
them 'materially impossible'.[69] This should begin with 'the entire
Franco–German production of coal and steel be[ing] placed under a common High Authority in an organization open to the interests of the other
countries of Europe'. Ending national competition in this strategically crucial industry would, Schuman argued, remove the principal cause of
Franco–German rivalry. Allowing other countries to join such an authority
would provide the 'common foundations for economic development as a
first step in the federation of Europe'.[70]

The fact that the Schuman Plan became official policy in France was
partly a result of domestic circumstances, notably the highly autonomous
nature of ministerial authority under the Fourth Republic which played into
the hands of small but dedicated lobbyists such as Monnet while allowing
'pet' projects to gather momentum inside particular ministries without
being undermined by rival politicians in other areas of government. But its
success also reflected the ability of Schuman and Monnet to promote their
plan amongst divergent, indeed logically opposing, constituencies. The idea
was self-evidently attractive to European federalists but it could also be
'sold' to traditional nationalists as an attempt to reassert an independent
European agenda which might eventually challenge the Anglo–American
alliance. This alliance was particularly worrying to French nationalist politicians. Schuman's proposal could be presented to his compatriots as a bid
for French hegemony in Europe (indeed it was partially informed by that
very ambition).[71] The idea could also be promoted amongst those who
sought an Americanised Europe. A successful common market in coal and
steel would modernise the key industrial sector and establish firm economic

foundations for a strong, western-oriented European pillar in an Atlantic alliance. This final argument proved persuasive to the US administration which endorsed the Schuman Plan, largely through the efforts of Monnet.[72]

By early 1951, five European governments (Belgium, West Germany, Italy, Luxembourg and the Netherlands) had accepted Schuman's invitation to take part in a conference to examine his plan. The United Kingdom rejected the idea, despite the enthusiasm of its own steel industry, partly on principle but also because a European arrangement was deemed likely to jeopardise British exports to the Commonwealth.[73] Complex negotiations ensued, with US blessing, about the degree of independence which the supranational agency should possess. The six governments then signed a treaty in Paris establishing the European Coal and Steel Community (ECSC) on 18 April 1951. This began operation the following summer with its principal architect, Monnet, in charge. The economic impact of the ECSC was considerable,[74] but its political implications are more relevant here. Although one cannot claim that its creation marked the beginning of an inexorable process towards European integration, the ECSC established the idea of an independent, supranational European authority with its own income (derived from an ability to tax a particular form of production) and its own judicial codes.[75]

Other European initiatives, also conceived in France at this time, were less successful. In the autumn of 1950, the French Prime Minister, René Pleven, sought to capitalise on the Schuman Plan by proposing a common European army and defence strategy. This was motivated by France's traditional fear of German rearmament, the inevitable consequence of a future West Germany being allowed to join NATO. A common European army would give France partial control over German military capacity. A draft treaty was duly signed by the six ECSC countries in May 1952 which looked forward to a European Defence Community (EDC) within NATO. Once again, the United Kingdom rejected the invitation to join.[76] Unlike the ECSC, which was clearly compatible with the idea of an 'Atlantic Europe' and with European national interests, the EDC seemed an unnecessary complication within NATO and also began to compromise the military independence of the European member states. Paradoxically, this latter problem was most acute in France itself, which was embroiled in an escalating imperial war in Indochina. Fearful that a fully operational EDC would restrict its military options beyond Europe, the French government refused to ratify the draft treaty in 1954, thus undermining an institution devised specifically to ensure French influence within Europe. A watered-down Western European Union (WEU) replaced the EDC in May 1955, an arrangement so innocuous that even the British signed up.

If the success of the ECSC demonstrated the potential of the European project in this period, the failure of the EDC reveals its limits. The lesson was clear: to be successful, pan-European institutions needed a pragmatic and primarily economic agenda. When the ECSC Council of Ministers met at Messina in June 1955 to discuss the prospects of further European integration, the schemes commanding widest support were those which focused on economic, technological and scientific co-operation, notably the idea of a wider common market amongst the 'Six' and the establishment of a European institution to co-ordinate atomic energy research and development.[77] Negotiations were by no means harmonious, however, particularly between France and the newly created Federal Republic of Germany, the former wishing to protect its agricultural sector from foreign competition, the latter seeking a free market to sustain its manufacturing sector. The uneven social costs of market integration and fears of rising regional disparities also generated tension within the 'Six', as did France's desperate rearguard attempt to retain an imperial, extra-European role in Algeria, where a bitter war raged from 1954 to 1962, and in Egypt, where Anglo–French forces were obliged to abandon their attempt to secure the Suez canal from Egyptian nationalisation in 1956 following US intervention. It was eventually agreed that a tariff-free market should be established between the 'Six', with standard external tariffs on imported goods and a common agricultural policy favouring France's large and still impoverished peasant electorate. Agreement was also reached on the complex matter of nuclear co-operation. These regulations were to be fully operational by the beginning of 1970 (though they were in place by 1968). On 25 March 1957, the 'Six' signed the Treaty of Rome creating the European Economic Community (EEC) and the European Atomic Energy Community (Euratom). These two agencies began operation the following January (Figure 4.5).[78] Although it was a recognisable offspring of the ECSC, the EEC revealed the shifts in the European debate during the 1950s. Whereas the Schuman Plan spoke of economic integration as a means of avoiding war, the preamble to the Treaty of Rome saw economic union as an end in itself.[79] The EEC offered the real prospect of a high-income, single European market. Once the threat of prohibitive external tariffs were removed, the new Community seemed, like its parent organisation, to be consistent both with an Atlantic, free-trade agenda and with the demands of national European governments.

The acceleration of the European project within the original 'Six' generated considerable unease in some sectors of the British political establishment. Involvement in supranational European integration, overseen by an unelected group of commissioners, commanded little support in the United Kingdom, but the prospect of total exclusion from Europe, an area increasingly important for Britain's trade, was even more disturbing. The Conservative Foreign Secretary, Selwyn Lloyd, attempted to 'parachute' Britain into the European project by announcing a 'Grand Design' in late 1955. This advocated bringing together the various European institutions in a central parliamentary structure which Britain would join. Those who were busy negotiating the terms of the Treaty of Rome understandably resented this intervention, which was also received without enthusiasm in

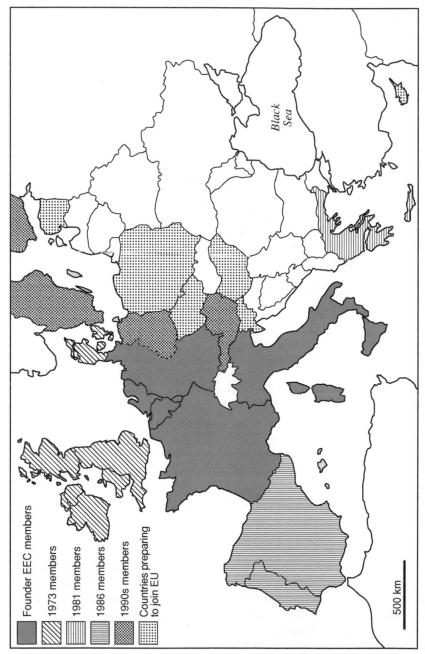

Figure 4.5 The emergence of the European Union, 1957–98.

Founder EEC members

1973 members

1981 members

1986 members

1990s members

Countries preparing
to join EU

*Black
Sea*

500 km

Washington.[80] Just over a year later, in February 1957, the United Kingdom relaunched its European policy with a proposal for a European Free Trade Association (EFTA) which would allow Britain, and any other countries which wished to join, the freedom to maintain their traditional non-European trading connections (particularly in foodstuffs, the staple colonial trading commodities) while gradually eliminating tariffs on industrial products within the association. This was a resolutely economic, inter-governmental and apolitical option which eschewed legally binding codes of practice and relied on the goodwill of its membership. Despite French opposition, seven non-EEC states (Austria, Denmark, Norway, Portugal, Sweden, Switzerland and the United Kingdom) signed the Stockholm Convention establishing EFTA in January 1960.

EFTA reflected Britain's 'minimalist' approach to European integration. It sought to stimulate trade between sovereign states and had no political or cultural aspirations beyond that limited objective. Within these terms of reference, it was a considerable success during the 1960s and early 1970s.[81] But the experience of establishing and organising EFTA produced quite different interpretations in the United Kingdom. Some saw it as a distinctive British route towards a more central European role, the seedbed of an Atlantic free-trade area in which Britain could act as the commercial bridge between the USA and western Europe. Others saw EFTA's economic success, which delivered higher overall growth rates amongst the 'Seven' than the 'Six' during the 1960s, as evidence of the dangers associated with a federalist programme. Hence, whereas Selwyn Lloyd insisted before the Council of Europe at the time of the Stockholm Convention that 'We [the British] regard ourselves as part of Europe, for reasons of sentiment, of history and geography', the Permanent Secretary at the British Treasury had penned an earlier memorandum warning against 'the kind of mysticism which appeals to European catholic federalists and occasionally – I fear – to our own Foreign Secretary'.[82]

The success of the European institutions created in the 1950s was determined by their ability to satisfy two criteria. First, they had to be reconciled with US Atlanticism and the wider Cold War geopolitical division established at the end of World War 2. Second, they depended on the active support of European governments and could not directly challenge the economic and political interests which these states believed they were protecting. Recent interpretations have emphasised this second geopolitical constraint. In contrast to functionalist and neo-functionalist theories of European integration which were constructed without the benefit of rich historical archives and which assumed that European institutions would seamlessly evolve towards a US-style federal system, research on unpublished official papers now in the public domain has produced a more inductive, empirical assessment of the early moves towards a European programme. Decision-making was less rational and more chaotic than functionalist accounts suggest and was determined largely by domestic interests.[83] Alan Milward has suggested that the process of European integration which began in the 1950s did not involve a marked break with traditional national geopolitics but rather a complex reconciliation of national and international objectives. There was no significant surrender of national sovereignty, he argues, but rather an enhancement of the power of traditional nation-states. Indeed, economic integration probably 'rescued' the western European nation-states from internal political upheavals (the likely outcome of the kind of economic stagnation which the OEEC, the ECSC and the EEC successfully avoided) and the very real threat of Soviet domination.[84] The fears of the latter-day 'Eurosceptics' and the hopes of the early 'Europhiles' were both misplaced, he implies, for the European project was determined by national requirements, draped in a European language:

> There is ... much to be said for the argument that the nation-state became more powerful after 1945 in western Europe than it had been before. It is oddly contradictory that theorists should have predicted the replacement of the nation-state at the exact time when European states were embarking on unprecedented programmes of intervention in economic and social life with the express purpose of shaping and controlling their national destinies.[85]

A corollary of Milward's argument is that the European programme was 'something made in Europe springing from the evolution of the European nation-state' rather than something imposed from without or driven by the relentless logic of Cold War geopolitics.[86] This is, to be sure, an argument which works most effectively for this early period of European integration and for the original 'Six' rather than the subsequent members of the European 'club', including the United Kingdom where the undoubted enhancement of state power after 1945 took place largely outside the main framework of European integration.

L'Europe des patries

The twin geopolitical constraints on the European institutions (their need to sustain a US Atlantic system and to support, rather than challenge, traditional nationalisms) created serious tensions during the 1960s and 1970s.[87] These were sparked off by attempts to enlarge the EEC from its original six members but rapidly developed into a more fundamental geopolitical dispute about the nature of Europe. The governments of the United Kingdom, Denmark and Eire applied to join the EEC in the summer of 1961. Harold Macmillan's Conservative government in London hoped to engineer a merger of the EEC and EFTA based on a future compromise between their different objectives. This was subsequently endorsed by the new US President, John F. Kennedy, in his Independence Day speech of 1962. This looked forward to 'a more perfect union' between the USA and Europe. But the prospect of British involvement in the EEC worried many in the United Kingdom and also disturbed a growing constituency on the continent. Some federalists were concerned that British involvement would act as a brake on the development of a strong and united Europe and would probably undermine any distant prospect of a pan-European system breaching the east–west divide. This reflected a growing fear that Europe was already too dominated by the USA, both culturally and geopolitically. In view of the linguistic and historical affinities between Britain and the USA, a prominent British role in the EEC would strengthen US influence and compromise European independence. This resulting tension was immediately exploited by the new Soviet leader, Nikita Khrushchev, who adopted a more conciliatory tone towards western Europe in the hope of driving a wedge between the EEC and the USA, explicitly referring in 1962 to the possibility of 'peaceful economic competition not only between individual states with different social systems but also between their economic federations'.[88]

In France, the concern that Britain would increase US domination fused with a more traditional, nationalist concern that the United Kingdom would undermine French hegemony within the emerging European system. The bitter memory of the 1956 Suez disaster (when British and French forces had been ignominiously forced to abandon their attempt to secure the Canal zone from Egyptian nationalisation following US pressure) still rankled in Paris, and US opposition to French plans to develop their own nuclear weapons programme made matters worse. The building of the Berlin Wall in the summer of 1961, though obviously a reflection of Soviet insecurity, seemed also to symbolise the definitive collapse of 'old Europe' and the triumph of the US geopolitics which appeared to have produced this demise. The anti-American lament was forcibly expressed in *Le Monde* on the first anniversary of NATO by Jean-Jacques Servan-Schreiber who argued that the Atlantic alliance had benefited the USA but had left Europe divided and weakened.[89]

French doubts about the Anglo–American axis and its enhanced role in Europe were embodied, in all their complexity and ambiguity, in the substantial form of the new French President, Charles de Gaulle, who returned to power, after years in the political wilderness, following a major constitutional crisis arising from the ongoing colonial war in Algeria. With a new Fifth Republic in place, de Gaulle launched his own nationalist vision of a *Europe des patries* which would stretch, he claimed, from 'the Atlantic to the Urals'. This was an audacious reassertion of an independent pan-European agenda, though it was based not on supranational institutions such as the EEC but on inter-governmental co-operation.[90] De Gaulle's vision was a kind of updated 'Concert of Europe' where economic and political agreements would be made in common but with the nation states still pre-eminent. He argued, for example, that a common European foreign and defence policy was possible, operating initially between the EEC membership but outside of its supranational conventions. He subsequently wrote:

[I]n order to achieve the unification of Europe', individual states are the only valid elements. . . . [W]hen their national interest is at stake nothing and nobody must be allowed to force their hands. . . . [C]o-operation between them is the only road that will lead anywhere. . . . What depths of illusion or prejudice would have to be plumbed in order to believe that European nations forged through long centuries by endless exertion and suffering, each with its own geography, history, language, traditions and institutions, could cease to be themselves and form a single entity?

Europe must never seek to model itself on the USA, a federation which had been created, claimed de Gaulle, 'from nothing in a completely new territory by successive waves of uprooted colonists'.[91]

Ironically, de Gaulle's hostility to supranationalism was broadly compatible with the attitude of successive British governments, but this narrow affinity was overwhelmed by his fear that Britain would destroy Europe's vulnerable independence and France's even more tenuous European hegemony. At a press conference early in 1963, following months of complex negotiations in which he had played no part, he tactlessly vetoed the British application. None of France's EEC partners had been informed and most were outraged both by the announcement itself and by the manner in which it was made. Goodwill, painstakingly developed over the previous years, quickly evaporated, despite a Franco–German Treaty of Friendship in early 1963 which obliged both governments to consult one another before making foreign policy decisions. The all-important EEC negotiations, particularly about the implementation of the Common Agricultural Policy, became increasingly fraught and matters came to a head in the summer of 1965 with the so-called 'empty-chair' episode when the French government, dissatisfied with the attitudes of its European partners, suspended all work within the EEC. The 'Luxembourg Compromise' of early 1966 overcame the immediate crisis but was based on an agreement to disagree rather than a genuine resolution of differences. The uncertainty was not helped by the confusion in Germany, where the new Christian Democrat leader, Ludwig Erhard (who had succeeded the long-serving Chancellor Konrad Adenauer) seemed unable to decide between a Gaullist or an Atlantic view of Europe and equally incapable of offering a German alternative.[92] The Common Agricultural Policy (a principal bone of contention which allowed, indeed encouraged, over-production through ensured, high prices) was finally agreed, though few regarded it as a definitive solution. Tensions did not end there, moreover, for French hostility to an Atlantic Europe resurfaced in the military arena. De Gaulle had previously refused to sign the 1963 Nuclear Test Ban Treaty and in February 1966 he announced his intention to withdraw from NATO on the grounds that France was not being accorded a sufficient role in the command structure. This was implemented in July, requiring, *inter alia*, the relocation of the organisation's headquarters from

Paris to Brussels. France's veto of a second application to join the EEC from Britain (now under a Labour government), Denmark and Eire in the summer of 1967 ended a difficult period which the near simultaneous merger of the ECSC, the EEC and Euratom into a single European Community (EC) did little to disguise.[93]

The future of the European project seemed extremely uncertain at the end of the 1960s. Attempts to create new European institutions were constrained by fundamental ideological divisions. The idea of an Atlantic Europe was challenged by a more independent, sometimes anti-American Europeanism, which occasionally looked forward to a reunification of the 'old Europe' which had been divided in 1945. Superimposed on this was a second disagreement between federalists, who imagined a new Europe replacing the old nation-states, and inter-governmentalists, who insisted on preserving traditional nation-states as the only viable building blocks of an integrated Europe. These difficulties were matched by growing attacks on the existing European nation-states 'from below'. A powerful, though ideologically incoherent form of oppositional politics developed at this time in which the young, in particular, voiced their dissatisfaction with the existing geopolitical order. As we have already mentioned, this 'grass-roots' desire for change manifested itself in Czechoslovakia during the 'Prague Spring' of 1968 before being brutally suppressed. In western Europe, similar impulses generated violent demonstrations in several cities against everything from US involvement in Vietnam to the structure of the university system. The student-led riots in Paris in May 1968 almost brought down the French Fifth Republic and paved the way for de Gaulle's resignation the following year.[94] According to Immanuel Wallerstein, 1968 was a:

> revolution in the world-system [which reflected] a deep skepticism about the liberal consensus that had dominated all the cultural, intellectual, and political institutions of the Western world. . . . What had been self-evident became debatable, and the group who had the right to enter the debate was no longer restricted to a small number of specialists.[95]

The late 1960s also witnessed the marked revival, and ideological transformation, of 'ethnic' nationalism which began to mount a serious challenge to the multi-ethnic nation-states of western Europe. Older, romantic and rather conservative forms of ethnic consciousness were gradually replaced by more radical, left-wing ethnic nationalism which, in its more extreme forms, endorsed the use of terrorism and violence.[96] Examples include the national separatist movements in the Basque area of northeastern Spain and on the island of Corsica in France. The mainly Catholic and republican minority in Northern Ireland launched a civil rights campaign against the widespread discrimination which they had endured under a provincial government dominated by Protestant Unionists, an attempt at peaceful change within the existing United Kingdom constitution which tragically descended into a cycle of violence that persists to this day. Elsewhere in the United Kingdom, in Scotland and Wales, nationalist political parties opposed to government from London enjoyed enhanced electoral support and reopened the debate about political devolution.[97]

The dilemmas of division

In West Germany the dissatisfaction with the existing geopolitical order, which had arisen in defiance of unprecedented levels of affluence in western Europe, provoked renewed debate about the possibility of 'healing the scar of Yalta', of restoring the integrity and cohesion of a divided continent. The belief that this might be possible was one outcome of a gradual thaw in the relationship between the USA and the USSR, itself a recognition of the crippling cost of nuclear proliferation in both empires. Willy Brandt, the socialist mayor of West Berlin, seized the opportunity of *détente* to launch a new strategy of *Ostpolitik*, first as Foreign Minister and subsequently as Chancellor.[98] This was a cautious attempt gradually to normalise relations between the two Germanys and, eventually, the two halves of Europe. Brandt's objectives were strictly limited and involved a step-by-step transition 'from conflict to co-existence to co-operation' ('*vom Gegeneinander über ein Nebeneinander zum Miteinander*'). The new policy also called for a relaxation in the division of Berlin which prevented even elderly relatives from visiting one another across the city. To avoid any impression that this was a prelude to German reunification, Brandt also sought a wider engagement with other eastern European states. Despite these modest aims, *Ostpolitik* required exceptional diplomacy for the idea, indeed the very word, conjured up the most alarming memories. Hard-line communists in the east suspected a scheme to uncouple East Germany from the Soviet bloc, 'aggression in carpet slippers', while some US commentators (notably Henry Kissinger) were also alarmed by this new show of West German assertiveness.[99]

Ostpolitik was an important development. Although its scope was limited by the enduring prerogatives of the Cold War it was based on an older idea of the integrity of traditional nation-states and of Europe as a whole, both east and west. It was also explicitly sustained by the familiar belief that Germany was a bridge between east and west, the heart of *Mitteleuropa*. Whether Brandt was motivated by a narrow German or a wider European concern is a difficult, and ultimately unanswerable, question but the positive aspects of his strategy were undeniably inspirational, notably his 'open-door' policy for immigrants to West Germany and his signing of the unilateral non-aggression accord with Moscow in August 1970. The latter was cemented by a famous visit to Warsaw the following December (to sign a similar accord with Poland) during which Brandt dropped to his knees before the memorial to the victims of Nazism in the city's former Jewish ghetto. *Ostpolitik*, and the wider relaxation of east–west tensions which it reflected and sustained, paved the way to the July 1973 Conference on Security and Co-operation in Europe (CSCE) which led to a complex era of east–west negotiations culminating with the Helsinki Final Act of the CSCE in mid-1975, the first major bridge across the post-war division of Europe. This confirmed Europe's borders as 'inviolable' and required NATO and the Warsaw Pact to provide advance warnings of troop manoeuvres and other military activities. The diminution in military tension facilitated the first stirrings of east–west economic contact, beginning with limited trade negotiations between the EC and Rumania in 1976 and with the USSR itself in 1977.

Ostpolitik was conceived by a national government rather than the EC and its modest success served to underline the inadequacy of the EC as a political forum. The momentum of western European integration faltered during the 1970s, in part because of global economic instability following the US decision to abandon the convertibility of the dollar into gold in August 1971 and in part because of the world depression, the result of a 1973 oil crisis which saw the price of crude petroleum quadruple then stagnate before rising sharply again in 1979–80. The conflict between federalism and inter-governmentalism continued within the 'Six', with opinion sharply divided between those who wished to 'deepen' the existing arrangements and move towards a fully-functioning federal system with its own budget and elected parliament, and those who wanted to enlarge the scope of the EC before progressing further. The former implied new federalist institutions; the latter implied more intergovernmental agencies. In the event, elements of both strategies can be detected after 1970.

The expansion of the EC to include the United Kingdom, Denmark and Eire from early 1973, following more than three years of complex negotiations, presented new challenges which delayed the 'deepening' of existing structures (Figure 4.5). First, the entry of the United Kingdom, which was still linked constitutionally to a sprawling 'Commonwealth' of former colonies in Africa and Asia, demanded a reconsideration of the EC's existing relationships with the developing world. The result was the Lomé Convention, signed in this west African city in February 1975. This allowed 46 African and Caribbean countries duty-free exports to the EC on most of their products.[100] Second, the largely urban and industrial nature of the British economy made the agricultural preoccupation of the EEC seem even more excessive. Partly as a result (though the initiative came mainly from Italy and Eire), a European Regional Development Fund was established in late 1974, with its own investment bank, to disburse EC funds towards the more depressed industrial regions and thus offset the economic disparities which seemed likely to increase as a result of European integration.[101] The Conservative government of Edward Heath, which negotiated Britain's entry, hoped that EC regional assistance might partially offset the enormous funds which the United Kingdom paid into the EC under the terms of its original membership.[102] But the election of Harold Wilson's more

'Eurosceptic' Labour government in 1974 led to a referendum on Britain's continuing EC membership. This produced a substantial 'yes' vote (67 per cent) but sufficient popular mandate for the United Kingdom to demand renegotiation of its terms of entry. Modest alterations were made but Britain's status as 'an awkward partner' was firmly established, throwing further doubt on the pace and direction of change. The UK government was broadly opposed to the federal position and strongly supported the new European Council which was established at the end of 1974, made up of the heads of the national governments. This reinforced the intergovernmental nature of the EC's institutional structure as did the likelihood of admittance of new states, notably the three Mediterranean countries of Greece, Portugal and Spain which had recently embraced democracy and sealed this with applications to the EC between 1975 and 1977.

The federalist ideal was far from moribund, however, and most EC agreements looked towards greater political unity. The Treaty of Luxembourg, signed in April 1970, set an ambitious course for a strengthened European Assembly and an increase in the EC's overall resources by 1975. At the Paris summit of October 1972, the EC heads of government agreed that economic and monetary union should be completed by the end of 1980, and the Copenhagen summit of December 1973 ended with a ringing Declaration on European Identity which looked forward to a common European foreign policy. Under the Presidency of Roy Jenkins (a former UK Chancellor of the Exchequer whose federalist leanings precipitated his departure from the British Labour Party), the European Commission sought to reinvigorate a federalist programme by establishing a European Monetary System (EMS) as the first step towards a common European currency, a serious alternative to the dollar. The EMS was designed to impose a measure of control over the value of EC currencies on the international markets and began to operate, after lengthy discussion, in the spring of 1979. In the same year, the populations of the nine EC member countries were allowed to elect their own representatives at the European Parliament in Strasbourg. These developments, more than the expansion of the EC, reflected the gradual shift from the idea of an Atlantic Europe which, if not dominated by US policy, was at least complementary to it, towards a more independent Europe which was capable of a co-ordinated economic challenge to the USA.[103]

Hope springs eternal

The advent of the 1980s offered little prospect of a dramatic change to the faltering progress of European integration. In eastern Europe, Soviet domination was as monolithic and unyielding as ever and, in the west, compromises between federalists and intergovernmentalists had produced a myriad of European institutions, few of which had any semblance of popular democratic accountability. Thirty-five years of endless negotiation had produced only 'Eurosclerosis', an institutional supranationalism which seemed distant, bureaucratic and vaguely sinister.[104] Although Greece became a member of the EC in 1981 and Spain and Portugal continued their preparations (both finally joined in 1986), it is revealing that Greenland became the first (and so far only) territory to withdraw from the EC following a referendum in February 1982 in the wake of the Common Fisheries Policy (Figure 4.5).

Those who hoped for a revival of Europe can scarcely have been encouraged by the election of 'new right' Conservative governments in the United Kingdom under Margaret Thatcher in 1979 and in the USA under Ronald Reagan in 1980. This established a new, aggressively Atlantic alliance based on shared ideological convictions. Both governments sought to reduce public expenditure by minimising the role of the state in the domestic economy and in social provision while simultaneously adopting a confrontational stance towards the Soviet bloc, 'the evil empire' in Reagan's preposterous language.[105] In the USA, this involved a massive expansion in military spending, a provocative and high-risk gamble designed, in part, to bring about an economic collapse in the USSR as it desperately sought to maintain parity. The ill-advised Soviet invasion of Afghanistan to quell a Muslim rebellion against the Marxist government intensified the east–west hostility. The world was plunged into the so-called 'Second Cold War' which seemed likely to destroy the hesitant attempts to build bridges across the ideological battlelines in Europe.

But while the early 1980s were indeed characterised by stasis and uncertainty, the middle and later years of the decade witnessed a dramatic return of serious political and intellectual debate about the meaning and scope of Europe, reviving discussions which had been moribund since Brandt's *Ostpolitik* in the early 1970s. Paradoxically, this was partly a result of the breakdown in the post-war liberal consensus which the new Atlantic alliance represented. The Anglo–American agenda was markedly at odds with the consensual orthodoxy of most western European countries, where economic and social policies remained interventionist and where geopolitical objectives were still influenced by the idea of *détente* with the east. The new conservatism was also entirely alien to the ideals represented by the EC. The election of François Mitterand as the French President in 1981 and Helmut Kohl as the German Chancellor the following year confirmed these very different European instincts, putting in place a Socialist–Christian Democrat alliance which became the backbone of the revived federalist programme. The ideological chasm between the Anglo–American alliance and the Franco–German perspective gave an added frisson to the rather stale debate about the future direction of Europe. At least four perspectives on the European project developed during the mid-1980s, each shaped by reference to each other. For convenience, these can be associated with the British, French, German and

Soviet governments, though it must be emphasised that this massively oversimplifies a more complex reality.

The first position, adopted by Thatcher and most of her close political allies, was anti-federalist and inter-governmental. Although the EC was hardly in a position to threaten national sovereignty in 1980, many British Conservatives were deeply concerned by the political predictions in the Treaty of Rome, though most were enthusiastic about the idea of a free-market, neo-liberal Europe. According to one British representative in the EC, from the outset Thatcher displayed 'a deep-seated prejudice' against the EC which she interpreted as a 'socialist' bureaucracy. Her worries were made worse by the apparent injustice of the United Kingdom's financial contribution to the EC, which far outweighed the immediate gains. The deficit stood at £947 million in 1979 and had risen sharply during the 1970s, despite the cosmetic adjustments following the 1975 renegotiation. The Common Agricultural Policy (which accounted for 70 per cent of EC expenditure) was the principal cause of British anger. This had little relevance for an industrialised economy with a small but highly efficient agricultural sector. As a substantial importer of food and manufactured products from beyond the EC, the United Kingdom also paid higher than average customs duties to Brussels. The early 1980s were dominated, therefore, by a tenacious British campaign to resolve the 'budgetary question' which effectively prevented further developments until a compromise was agreed in the summer of 1984.[106] Once this matter was settled, Thatcher sought to change the nature of the EC to reflect better her ideological and geopolitical convictions. Apart from her strongly Atlantic preoccupations, Thatcher's view of Europe bears some comparison with de Gaulle's *Europe des patries*. Like de Gaulle, she argued for an expansive but 'shallow' Europe, a structure which embraced as many countries as possible, preserved their national independence and limited itself to encouraging commerce and trade. The Thatcherite view was famously rehearsed in a speech at the College of Europe in Bruges on 20 September 1988. Here she spoke of a 'family of nations' coming together in an intergovernmental forum. The 'willing and active co-operation between independent sovereign states is the best way to build a successful European Community', she argued. Like de Gaulle, she explicitly rejected the federalist idea of a United States of Europe, based on the US model. Europe must reject protectionism, she further insisted, and encourage enterprise. More specifically, its common policies (particularly the Common Agricultural Policy) must be completely reformed.[107] To some extent, this viewpoint was echoed by influential commentators in Washington who were now convinced that the EC was no longer compatible with US economic or geopolitical interests. Samuel Huntington, an influential voice on US foreign relations, believed that '[t]he political integration of the European Community, if that should occur, would bring into existence an extraordinarily powerful entity which could not help but be perceived as a major threat to American interests'. The USA should therefore promote the 'evolution of the European Community in the direction of a looser, purely economic entity with broader membership rather than a tighter political entity with an integrated foreign policy'.[108]

Countering this view was the orthodox federalist position, which ultimately triumphed in Paris. This was expressed, in its purest form, during the summer of 1980 by the former Italian resistance leader and veteran European campaigner, Altiero Spinelli, who proposed a radical plan to increase the EC's democratic legitimacy by transferring executive power from the Commission and the Council of Ministers to the European Parliament within a reformed European Union. Spinelli's scheme was designed to remove the Eurosceptic criticism that a federalist Europe was inherently anti-democratic. It was adopted in spirit in socialist France where Mitterand was anxious to develop a distinctive, post-Gaullist European policy.[109] The inter-governmentalism which dominated French attitudes in the 1960s and, to a lesser extent, the 1970s was gradually replaced by a policy broadly comparable with that advanced by Monnet and Schumann in the late 1950s. The old, Napoleonic equation of France with Europe was reasserted, along with the pre-Gaullist argument that a federal Europe was the natural culmination of France's national destiny. A new Europe meant a new and expanded France.[110] This implied a fundamental dispute with the inter-governmental view of political sovereignty. While the British constitution was built on the idea of shared cultural identity which allowed individual subjects to profess themselves to be either English, Welsh, Scottish or Northern Irish as well as British, this did not extend to political sovereignty which resided only with the national Parliament. Thatcher interpreted political sovereignty in all-or nothing terms; you were either British or you were not. The idea that sovereignty might be shared between two or more nation-states or even 'pooled' within a supranational European federation was anathema. Yet this was precisely what federalists sought to achieve. Sovereignty, they argued, was endlessly flexible and no longer limited by traditional territorial jurisdictions. The tension between inter-governmental and federal viewpoints was such that some commentators began to contemplate a 'two-speed Europe', a 'variable geometry' in Mitterand's 1984 phrase.[111] Neither side relished this prospect, however, for it seemed to herald the unravelling of the European programme. Just over a year after Margaret Thatcher's speech at Bruges, the French President of the European Commission, Jacques Delors, set forth a federalist alternative to her view at exactly the same venue. An inter-governmental arrangement would never survive the vicissitudes of a global economy and would at some point break apart, he argued, leaving the peoples of Europe vulnerable and isolated from one another. However, the fear that a federal Europe would ride roughshod over national or regional peculiarities and governmental structures was misplaced. Federalism and inter-governmentalism could be reconciled without recourse to a 'two-speed Europe' through the concept of 'subsidiarity', pithily summed up by the adage 'Never entrust to a bigger unit anything which

is best done by a smaller one'. Wherever possible, decision-making should
be taken at regional or national level or, as it was expressed later, 'as closely
as possible to the citizen'. The purpose of the EC was to provide a loose,
overarching structure within which these decisions could be taken. A federal
Europe did not imply an homogenous, centralised Europe for it was based
upon the very opposite ideals of pluralism, decentralisation and diversity.[112]

This straightforward dispute between inter-governmental and federal
viewpoints was complicated by the (re-)emergence of a third, central
European perspective associated with German federalists and dissident
intellectuals in communist states such as Poland, Czechoslovakia and
Hungary. Inspired by the heroic campaign for political and economic
reform in Poland (organised by Solidarity, a trade union led by Lech
Walesa, a Gdansk shipyard worker), ideas which had lain dormant for years
resurfaced with a vengeance. Prominent amongst them was the familiar concept of *Mitteleuropa*. In contrast to the cautious approach adopted by
Brandt, a more expansive and optimistic debate took place after 1985 about
the possibility of breaching the ideological divide established in 1945 and
creating a new, cosmopolitan *Mitteleuropa*, united by its very diversity. This
was an old idea, of course, but was expressed in the new, 'postmodern' language of cultural hybridity and multiple citizenship.[113] The outstanding
statement came in a prophetic 1984 essay by the Czech writer Milan
Kundera in *The New York Review of Books*. This was a powerful reassertion of the 'Europeanness' of those living under communist governments
beyond the 'iron curtain'. The idea that a new European identity might
emanate from institutions devised by an unrepresentative technocratic élite
in Brussels who had little contact with the peoples of western Europe and
none whatsoever with the inhabitants of central Europe was manifestly
absurd, Kundera implied. A 'Europe' cut off from such quintessentially
European cities as Budapest, Prague, Leipzig and Warsaw made no sense.
Europe could never exist without a new *Mitteleuropa* at its core.[114] A similar point was made the following year by Hugh Seton-Watson in one of his
last essays. Despite being sealed off under Soviet domination since 1945,
central Europe had produced the most articulate and passionate 'pro-
European' liberals, he argued: 'Nowhere in the world is there so widespread
a belief in the reality, and the importance, of a European cultural community, as in the countries lying between the EEC and the Soviet Union'.[115]

The prospect of a revived *Mitteleuropa*, however far off this must have
seemed in the mid-1980s, raised obvious fears across Europe, and within
Germany itself. The historical associations of this term were still very much
alive. The strongly federalist line adopted by Helmut Kohl during the 1980s
needs to be seen in this context for this was a consequence of Germany's
position athwart a previously frozen ideological divide which was beginning
to show the first vague stirrings of movement. Kohl's predecessor as
Chancellor, Kurt-George Kiesinger, memorably described Germany's
dilemma in 1967: 'an awkward size in an awkward place ... too big to play

no role in the balance of forces ... too small to keep the forces around it in balance'. This view was echoed even more succinctly by Henry Kissinger who believed Germany 'too big for Europe, too small for the world'.[116] This meant that while France needed the EC to spread its influence across Europe and the wider world, Germany needed Europe to suppress its 'natural' tendency to dominate the continent under the old form of national geopolitics. It is for this reason that a senior German diplomat claimed that his foreign policy decisions were guided by the paradoxical aim of preventing German hegemony.[117] As Kohl himself put it in April 1990, 'The first important principle is that Germany must not be singled out ... [This] will turn the geographic centre of Europe into a ghetto – a policy which would be catastrophic'.[118]

The fourth perspective on Europe emerged after 1985 in the USSR under the restless and reforming leadership of Mikhail Gorbachev.[119] The rise of Gorbachev was itself evidence that the 'Second Cold War' was placing unbearable burdens on the eastern alliance and on the economic system of the USSR. The myth of Soviet military invincibility had already been undermined by its failure to control the rebellion in Afghanistan, and reform-minded communists, led by Gorbachev, were convinced that the reduction of international tension was crucial to the very survival of the Soviet empire. This required a radical approach on four fronts. First, the immense financial burdens of maintaining the Soviet military capacity, particularly its nuclear arsenal, needed to be dramatically reduced. Gorbachev initiated this policy almost immediately, notably at the Reykjavik summit with Reagan in September 1986 when his proposal to remove most nuclear weapons from European soil surprised the Americans and was consequently rejected, only to be revived in muted form the following year with the agreement to dismantle all Europe-based intermediate-range nuclear warheads. One implication of this rush to disarm was a revival of interest in the Western European Union, the specifically European military alliance within NATO which had replaced the European Defence Community in the mid-1950s but had otherwise lain dormant. Second, a new domestic policy of economic restructuring was rushed into place under the name *perestroika*, a term Gorbachev used as the title for his 1987 book. The objective was to relax the iron grip of communist economic planning and introduce a measure of free enterprise. This was linked to a third policy of social, cultural and intellectual openness, or *glasnost*, involving the removal of censorship laws and many of the other restrictions on personal freedom. Fourth, and most importantly here, Gorbachev proposed a new European policy based on the rallying cry of a 'Common European Home'. The USSR had no aggressive intentions towards Europe, he insisted, for the Soviet republics were an integral part of this international arena. This represented a dramatic shift in Soviet rhetoric which had previously avoided reference to Europe except as a narrow, descriptive term. The process of western European integration had been denounced as an instrument of capitalists seeking to shore up their faltering authority by facilitating international trade and intensifying their exploitation of the working classes. Gorbachev's new language drew upon, and sustained, the *Mitteleuropa* arguments developed in Germany and central Europe (see above). Like these related debates, Gorbachev's insistence that Russia was historically, culturally and politically European was a restatement of a much older view, in this case the pro-western Russian perspective which, as we have seen in earlier chapters, can be traced back to Peter the Great. Gorbachev's case for the USSR's 'Europeanness' drew upon some decidedly 'revisionist' Marxism. In his volume on *Perestroika*, he returned to an argument which the Duc de Sully had refuted over three centuries earlier. The USSR was part of Europe, Gorbachev claimed, because it was a Christian country which had, moreover, saved the rest of Europe from the threat of Nazism. Although he did not explicitly deny the right of the USA to intervene in European matters, this was initially the subtext of Gorbachev's pronouncements. An Atlantic Europe, wedded to the USA and excluding the eastern European states, was 'a chariot of war ... loaded with nuclear explosive'.[120] The acceptance of the Soviet Union as a part of the 'Common European Home' was, on the other hand, the first step towards permanent peace. As Neil Malcolm argues, however, Gorbachev's policy was probably motivated by a recognition that Soviet economic and military domination was on the wane and needed to be bolstered by a more subtle cultural and political presence in Europe.[121] Whatever his motives, Gorbachev's policy immediately affected the relationship between the EC and COMECON, leading to a Joint Declaration on 24 June 1988 in which the two organisations agreed to 'develop co-operation in areas which fall in their respective spheres of competence and where there is a common interest'. This accord also allowed individual COMECON countries to negotiate freely with the EC, effectively undermining the principal discipline which the former organisation had previously maintained.[122]

The developments within the EC during the mid-1980s reflected these four different visions of Europe as well as the new opportunities offered by the upturn in the world economy. The conflict between the federal instincts of some EC states (notably France and Germany) and the intergovernmentalism of others (particularly Britain) was avoided by the familiar expediency of allowing both sides to promote different schemes within the EC. The federalist view was reinforced by a Solemn Declaration of European Unity, which the EC heads of state signed in the summer of 1983, and by the European Parliament's adoption in principle of a draft treaty on European Union (a modification of Spinelli's earlier proposal) in February 1984. The intergovernmental interests of the British, and their desire to widen rather than deepen the EC, was also sustained by the agreement early in 1984 to establish a common free-trade zone encompassing the EC and the remaining EFTA states (Austria, Switzerland and the Scandinavian countries) and by the aforementioned engagement with COMECON. These expansive economic strategies were designed to counteract growing fears in the USA and around the world that the EC was moving towards economic autarky, a 'fortress Europe'.

Bringing together the jumble of existing economic agreements into a single act to complete the common market, the ideal enshrined in the Treaty of Rome, was a strategy which appealed both to federalists and intergovernmentalists (although a significant minority on both sides worried that without a stronger regional policy a completely integrated market would exacerbate geographical inequalities). This rare consensus allowed the Single European Act (SEA) to be agreed with uncharacteristic despatch. Delors's White Paper, *Completing the Internal Market*, was drawn up in the spring of 1985 and recommended over 300 measures to be implemented by the end of 1992. The SEA committed the EC to establishing an integrated market across all sectors of the economy, 'an area without internal frontiers in which the free movements of goods, persons, services and capital is ensured'.[123] The plan was formally adopted by the European Council at the end of 1985 and came into operation in July 1987.

The consensus over the SEA was more apparent than real, however, for the idea meant different things to federalists and intergovernmentalists. The former saw it as another step on the road to political unity, the latter as the terminus. Encouraged by the illusion of unity, however, Delors pushed ahead with plans for European Monetary Union (EMU), building upon the European Monetary System introduced by Roy Jenkins in the late 1970s. His three-stage plan, published in April 1989, looked forward to a common European currency, the 'logical corollary' of the SEA, according to Delors. The first stage was sustained convergence of all EC currencies through the co-ordinated action of the national central banks operating under the terms of the European Exchange Rate Mechanism. The second stage would transfer the management of monetary policy from national governments to a network of national central banks which would be required to act in unison to keep inflation, the money supply and interest rates within strict limits. The third phase would begin with the locking together of the different national exchanges into a common system. Two months later, in June 1989, the European Council agreed that the first stage of the Delors Plan should begin the following July. Needless to say, EMU represented a federalist leap which Margaret Thatcher and her supporters (who were by no means limited to the United Kingdom) were unable to accept. A crisis erupted in the UK government which ended with Thatcher's resignation in 1990. In the midst of these events, Turkey and Austria applied for EC membership, though the proposal from the former country was rejected, partly because of the government's lamentable civil rights record.

The sudden acceleration of the western European project during the late 1980s has provoked extensive debate. The changes taking place in the Soviet Union and eastern Europe were clearly important, though their precise relationship to events in the EC is by no means clear even now. Some scholars insist that although Thatcher was a notable casualty of the revived

European programme, the new momentum was still based on domestic, national calculations and was therefore an intergovernmental rather than a truly federal development. Detailed studies of the negotiations themselves suggest that the traditional national concerns still dominated. Despite the fears of Eurosceptics, the SEA and EMU were primarily economic agreements, the lowest common denominator in a complex process of international discussion.[124] According to Milward and Sørensen, for example, writing in 1993, the SEA and EMU did not fundamentally challenge the power of the traditional nation-state: 'The frontier of national sovereignty ... remains with little alteration where it was fixed in ... 1957'.[125] Others suggest that this fails to recognise the influence of 'non-political' actors and organisations, operating away from the negotiating tables of the EC and beyond national governments. According to Sandholtz and Zysman, the European revival after 1984 was triggered by European business interests and corporate élites who supported a common market and a single currency and successfully lobbied both national governments and the EC itself.[126]

The sadness of geography[127]

These developments, critically important in the context of western European integration, were overshadowed by the extraordinary events in east-central Europe during the last weeks of 1989. The political landscape of this region, frozen for a generation, was utterly transformed in a few short days following the unexpected withdrawal of Soviet support for the communist regimes in the satellite states. These collapsed one after the other, accompanied by rather less violence than had initially been feared. The Warsaw Pact and COMECON, the institutions through which Soviet power had been exercised in Europe, simply dissolved (though they were not formally disbanded until 1991). Suddenly the misleading short-hand terms of Cold War geopolitics – 'east' and 'west', 'first world' and 'second world' – made no sense. Faced with the mass exodus of East Germans to the Federal Republic, where they were constitutionally guaranteed citizenship, the communist authorities in East Berlin opened the wall. The ultimate symbol of the Cold War division of Europe crumbled into a mass of souvenirs, hacked away by ecstatic, pick-axe-wielding crowds before the astonished eyes of the millions watching on television. German reunification, a far-off dream only days before, became a reality virtually overnight. Ignoring the serious reservations of his European partners, particularly France and Britain, Kohl seized the opportunity of reunification with alacrity, dismissing fears about the enormous costs and technical difficulties associated by such a merger, particularly the creation of a single currency across the new state. A 10-point plan was announced on 28 November 1990 which studiously avoided all mention of the word 'unification', though no one had the slightest doubt about what was happening. The USA offered its endorsement and in February 1990 the USSR recognised the *fait accompli* of a united Germany and, moreover, its place within NATO. Formal unification took place the following October, thus expanding the EC to incorporate the area of the former Democratic German Republic. Eight months later, in June 1991, the German Parliament took the powerfully symbolic decision to re-establish Berlin as the new German capital.[128]

German unification and the collapse of communism generated wild enthusiasm about the prospects of a reborn, reunited Europe. Even those who were worried by German unification were swept along by the emotion of the times. In a memorable New Year's message to the French people in January 1990, François Mitterand summed up the general mood: 'Europe is returning to its history and its geography'. By August 1991 (in the midst of preparing an unprecedented alliance to 'liberate' the oil fiefdom of Kuwait from its Iraqi conquerors, with the support of the USSR), the US President George Bush foresaw 'an historic period of co-operation . . . a new world order' in which the old ideological divisions between 'east' and 'west' no longer applied.[129] The farcical attempted coup by old-style communists in the USSR a month later sent shock waves around the world but its immediate failure simply undermined the last vestiges of communist authority in the USSR and handed power to Gorbachev's post-communist opponent, Boris Yeltsin. The USSR, an awesome superpower a few months earlier, ignominiously collapsed as its constituent republics broke away and declared their independence of Moscow. Similar centrifugal forces had already begun to dismember the other multi-ethnic federations of central Europe. Some 14 new states eventually appeared on the European map, creating a degree of geopolitical complexity not seen since the eighteenth century (Figure 4.6).[130] In most cases, this took place more peacefully than anyone imagined possible. The Baltic States, Belarus, Ukraine and Moldova arose from the ashes of the former USSR with little bloodshed, and the dismemberment of Czechoslovakia into separate Czech and Slovak Republics in early 1993 was described as a 'velvet divorce', the 'logical' extension of the collapse of communism. A thousand, half-forgotten symbols of nationhood resurfaced across the continent as if by magic. Flags that had been kept hidden for decades fluttered proudly on rooftops and street lights; anthems that had been sung in hushed tones in smoky back rooms were roared confidently by great celebratory crowds; stories of wars and battles, of the heroes and heroines of nationalist struggles, became the mainstay of reformed educational systems; while ancient maps of new national spaces, the cartographic 'proof' of historic legitimacy, were reissued in expensive facsimile editions and dispatched around the world.

Conservative commentators in the west were openly triumphalist. Francis Fukuyama saw the collapse of communism as a straightforward victory for liberal democracy, the 'west' defeating the 'east' in the war of ideas. History, in the sense of an ideological conflict between different worldviews, had come to an end. Free peoples all over the world were now able to

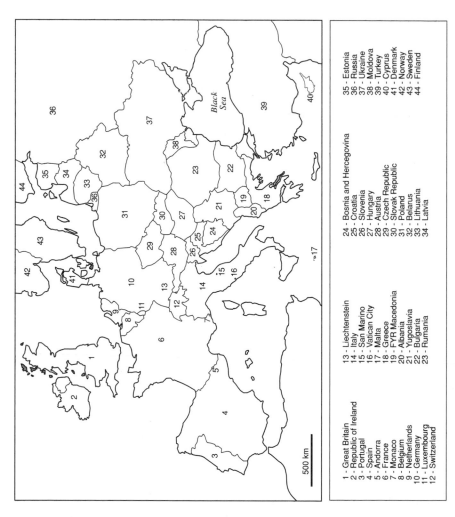

1 - Great Britain
2 - Republic of Ireland
3 - Portugal
4 - Spain
5 - Andorra
6 - France
7 - Monaco
8 - Belgium
9 - Netherlands
10 - Germany
11 - Luxembourg
12 - Switzerland

13 - Liechtenstein
14 - Italy
15 - San Marino
16 - Vatican City
17 - Malta
18 - Greece
19 - FYR Macedonia
20 - Albania
21 - Yugoslavia
22 - Bulgaria
23 - Rumania

24 - Bosnia and Hercegovina
25 - Croatia
26 - Slovenia
27 - Hungary
28 - Austria
29 - Czech Republic
30 - Slovak Republic
31 - Poland
32 - Belarus
33 - Lithuania
34 - Latvia

35 - Estonia
36 - Russia
37 - Ukraine
38 - Moldova
39 - Turkey
40 - Cyprus
41 - Denmark
42 - Norway
43 - Sweden
44 - Finland

Figure 4.6 A united Europe? The political geography of post Cold War Europe.

re-establish their personal and collective identities within a new global consensus.[131] Duignan and Gann were even more upbeat, writing of the 'stun-
ning victory' which the US and its 'democratic allies' had achieved over
'communism, fascism and other authoritarian systems':

> The Atlantic partnership born of World War II has since shaped the
> modern world. It was a partnership that created peace and prosperity
> for Western Europe and North America alike, after many years of tur-
> moil and bloodshed. It was a partnership which defeated the Soviet
> challenge and ended the Cold War. It is this partnership which the
> newly independent countries of East-Central Europe wish to join in
> some form. It is this partnership which has created innumerable link-
> ages, official and non-official, uniting the North Atlantic states – the
> most successful alliance in modern history.[132]

As the euphoria subsided, however, darker fears began to surface about
the new European order. A more pessimistic reading of 1989 saw it not as a
positive triumph of liberal democracy, the new 'springtime of nations', 1789
and 1848 rolled into one, but rather as a negative rejection of an unworkable and inefficient ideology. The transformations of 1989 seemed to be neither the outcome, nor the source, of new social, moral or ethical
convictions.[133] This was the ultimate contradiction in terms: a conservative
revolution, a confused throwback to a pre-communist past reflected in the
changing language of everyday life. In the former German Democratic
Republic, for example, communist terms such as *comrade* became outdated
almost immediately only to be replaced by even older expressions, such as
Burger, which suddenly acquired a modish appeal. The relative absence of
violence in 1989 also suggested meek surrender, the apologetic acceptance
that, on balance, western-style capitalism delivered higher living standards
and a better quality of life than Soviet-style communism (which was, of
course, the only style of communism on offer). The relentless materialism of
the west was not the price worth paying for greater freedom, it was the prize
itself. Greater freedom was simply an added bonus. The result was an unin-
spiring 'colonisation' of central and eastern Europe by a recharged and
rapacious consumer capitalism, as culturally depressing as it was threatening to the still fragile democracies of this region. Societies which had been
spared the dubious post-war 'pleasures' of fast-food, satellite television and
computer pornography suddenly found themselves overwhelmed by these
and other tawdry aspects of late twentieth-century western life. The imposing statues of Marx and Lenin had crashed to the ground only to be replaced
by the neon lights of all-night garages and the golden archways of another
MacDonalds restaurant.[134] Neither could one point to the unbounded joy of
those on the receiving end of this invasion. The new freedom seemed to
bring only intensifying economic crisis compounded by further environmental disasters, including the calamitous flooding of the river Oder along
the German–Polish border. Is it any wonder, then, that Gören Therborn

should find that 'Eastern Europeans were in 1991 the unhappiest peoples on the earth'?[135]

While no one mourned the passing of repressive and totalitarian regimes (except those who had directed them), the break-up of multi-ethnic federal states into new nations based on ethnic identities raised legitimate fears that 1989 had reflected and encouraged a racial, xenophobic geopolitics. This was, of course, the same argument which raged during World War 1 about self-determination and 'Balkanisation', an historical parallel which led Eric Hobsbawm to describe the explosion of national separatism after 1989 as 'the unfinished business of 1918–21 ... the old chickens of Versailles ... coming home to roost'.[136] The ironies were obvious: the 1989 revolutions, loudly proclaimed as the beginning of a reborn Europe in which the east would be drawn together with an increasingly integrated west in a centripetal process, had generated only a centrifugal blast of national separatism. Just as other, supposedly less 'developed' parts of the world were removing the outdated geopolitics of ethnic separatism (most obviously in South Africa), Europe was busily reinventing it.[137] On a more prosaic but no less important level, the sheer complexity of the new European order seemed to make the system less predictable and more prone to localised conflict. The fact that nuclear weapons, previously controlled by a single superpower, were now scattered across several, much smaller and independent states was particularly worrying.[138] The immense economic difficulties confronting the new states also raised fears that eastern Europeans might end up rejecting both communism and liberal democracy, paving the way for the rebirth of fascism.[139] The disturbing rise of racism and neo-fascism during the early and mid-1990s in Germany and Russia, involving murderous attacks on Turkish immigrants and the desecration of Jewish cemeteries, suggested such fears are not entirely misplaced.[140]

But to view the new nationalisms of eastern and central Europe as a return to the past ignores the thoroughly (post)modern nature of the revolutions which created them. The novel features include a millennial environmentalist critique that reflected the appalling record of ecological damage under Soviet domination and which intensified in the wake of the Chernobyl disaster of 1986. The role of the western mass media was also important. The revolution in information technology in the west during the 1980s made it impossible to maintain control over the flow of information in eastern Europe. Gorbachev's policy of *glasnost* was partly an acknowledgement of that fact. The long-established media of radio and terrestrial television were joined by a host of other technologies which facilitated the dissemination and reproduction of western cultural norms: citizens' band (CB) radio, cheap computers and photocopiers, tape recorders and videos (with pirated tapes), mobile phones and illicit satellite television receptors. These were not conduits for elevated alternative ideologies, of course, but they served to demonstrate the yawning chasm between life on either side of the ideological divide. The apocryphal joke was that cats in the west were seen enjoying a more nutritious diet than people in the east. As Pearson notes, 'The social role of television [in eastern Europe] switched from a reliable anaesthetic to an uncontrollable stimulant'.[141] The onset of a deterritorialised, 'cyberspace culture' both undermined and overwhelmed the traditional earthbound, territorial authority of communist rule, symbolised by the Berlin Wall.[142] This despised structure, so imposing in 1961, seemed almost moribund by 1988. The irony, of course, is that while a deterritorialised, global media helped to destroy communism, a brutally integrative territorial system under the USSR, it paved the way for a reterritorialised, ethnic nationalism in its place. Globalisation and the creation of entrenched ethnic identities were thus directly related: a new 'civilisation of the satellite' produced a new 'civilisation of the soil'.[143]

The latter reached its apotheosis with the outbreak of war in the disintegrating Yugoslavia in the autumn of 1991, following months of skirmishing between rival ethnic groups. The tragedy of Yugoslavia, and more particularly Bosnia, demonstrates the enduring potency of European territoriality. Although it is all too easy (and often necessary) to apportion blame amongst the warring factions and their leaders, the temptation to see the ongoing Yugoslavian crisis as visceral tribalism, a barbaric and primitive form of ethnic hatred which was merely held in check under communism only to be unleashed when these 'backward' peoples were at last free to massacre one another, is both intellectually lazy and morally complacent.[144] As David Campbell demonstrates in his brilliant analysis, Yugoslavia lies at the heart of Europe. It is populated by sophisticated, highly educated, thoroughly modern inhabitants. It is not over there; it is here. Its problems are the problems of Europe. As Campbell shows, even before the crisis descended into the fury of warfare and ethnic bloodletting, the 'international community' had privileged a territorial solution above all other alternatives.[145] The terms of the debate were set in advance by enduring assumptions that national identity could only be expressed by authority over space. One was confronted by the strange paradox of politicians from western European countries who had spent most of their adult lives working towards a federal, cosmopolitan and borderless European Union enthusiastically supporting the break-up of Yugoslavia into a mass of ethnically cohesive, strictly delimited states. This was particularly noticeable in Germany where Kohl's government, still flushed by its own triumphant national self-determination, bounced the rest of the EC into recognising the independence of Slovenia and Croatia in the summer of 1991. While a territorial solution was at least feasible in the case of Slovenia, it condemned multi-ethnic, cosmopolitan Bosnia to a savage death of territorial attrition and attempted genocide. Most of the federal republics were ethnically mixed but Bosnia, in the centre of the old federation, was a bewildering mosaic of different communities in the rural areas. In the capital city of Sarajevo, Orthodox Serbs, Catholic Croats and Bosnian Muslims fused in a seamless mixture and had lived in peace for generations. A neat territorial solution would never be possible

here; Bosnia's fate was sealed by the desire to sanction such a solution elsewhere. The quest for ethnic homogeneity simply created the worst kind of
heterogeneity: small, isolated and vulnerable minorities beyond each new
suggested border, the targets for those who were willing forcibly to relocate
people (or worse), the practice which acquired the absurd euphemism of
'ethnic cleansing'. The belief in the inalienable right of the former
Yugoslavian republics to secede and the failure to offer any kind of moral or
intellectual defence of a non-territorial alternative to the resulting crisis
made territorial conflict and 'ethnic cleansing' almost inevitable. The continual attempt to establish a neat cartographic solution (the dominant
objective from the Lisbon accords in March 1992 through the Vance–Owen
Proposals to the Croat–Muslim Federation and the Contact Plans of 1994)
played directly into the hands of those in all communities, though particularly amongst the Serbs, who were itching to seize as much territory as possible by force (Figure 4.7).[146] Despite a half century of European debate in
which sovereignty and citizenship seemed to be increasingly uncoupled from
land and territory, the Yugoslavian crisis demonstrates the remarkable persistence of older forms of geopolitical reasoning both amongst the participants in the war and amongst those who sought to arbitrate. Herein lies the
'sadness of geography'.

The EC, an organisation supposedly devoted to breaking down borders
between European peoples and which had already welcomed former communist countries onto the Council of Europe, manifestly failed to provide
any kind of political leadership in Yugoslavia.[147] Despite fears that the EC
had become too political, the Yugoslavian crisis demonstrated it was still
primarily an economic organisation. No coherent common policy was
devised with respect to Yugoslavia, the different EC nation-states responding in quite separate and rather traditional ways, revealing thereby the
weakness of the supposedly 'indissoluble' bonds between them. And yet,
even as the war intensified, the familiar, by now rather depressing, arguments about the nature and pace of European economic integration continued. The federalist agenda was hugely reinforced by a series of
intergovernmental conferences on EMU and political union which began in
December 1990 and culminated with the signing of the Maastricht Treaty a
year later. This was a rather messy and ambiguous agreement, not least
because the United Kingdom insisted on the right to 'opt out' of the social
chapter which it regarded as a 'socialist tax on jobs' which would end up
increasing unemployment not only in Britain but across the EC. Despite
this, the Treaty on European Union committed the EC to a form of integration noticeably more political than anything before. The central difference
lay in monetary union and the inevitable loss of national control over currencies, still seen as the cornerstones of state sovereignty. The necessary ratification of the treaty at the national level was the biggest test yet of the EC's
ability to convince the European peoples that their interests lay in greater
unity. There was no mechanism to standardise the procedure used for

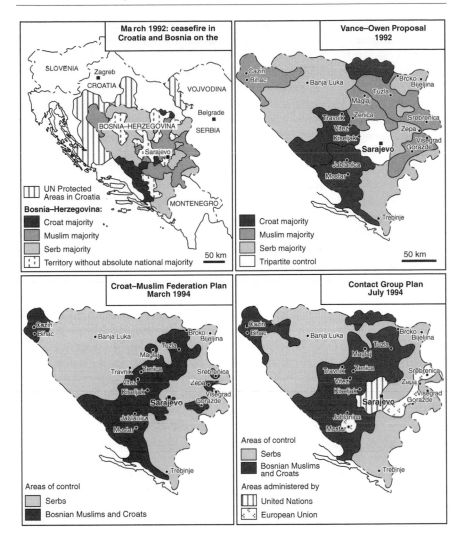

Figure 4.7 *Plus ça change?* Mapping Yugoslavia in the 1990s. Source: after Popovski (1995, 202); Silber and Little (1996, viii–xi).

ratification and some countries (including the United Kingdom) relied on a simple parliamentary majority while others (including France) organised a referendum. Interestingly, the latter course produced the major difficulties. The Danish population narrowly rejected the treaty in June 1992 only to change its mind in a second ballot the following May, while the French accepted it by the narrowest of margins in September 1992. The UK Parliament eventually ratified the treaty the following July, having reached final agreement over its 'opt-out' clause the previous December, to the continuing dismay of many.[148] With the single internal market now in place and the ratification underway, the EC was formally renamed the European Union in November 1993.

The timetable for this transition remains a matter of intense dispute (which was notably heightened after September 1992 when both Britain and Italy were forced to withdraw from the ERM, the system which sought to discipline European currency fluctuations). At the time of writing, few of the EU nations meet the so-called 'convergence criteria' laid down in the Maastricht Treaty as necessary for a single European currency to come into existence as anticipated, by the end of the century. The 'deepening' of the EU into a genuinely federal system is also challenged by the pressures to enlarge. The admittance of Sweden and Finland during the early 1990s created an unprecented north–south expansion, a Europe stretching from the Canary Islands to Lapland. The biggest challenge, however, comes from the east. Hitherto, the EC/EU's attempts to create institutional bridges with the former communist states have failed. The fiasco surrounding the establishment of the new European Bank for Reconstruction and Development in 1991 to direct EC funds to eastern Europe is illustrative. Despite its pan-European rhetoric, this was a French idea and seems to have been partially motivated by a traditional nationalist fear that without a European (read French) role in the east, this whole arena would become a German zone.[149] Despite such débâcles, there are now 11 former communist states which have expressed a desire to join the 15 EU members.[150] Representatives of all 26 nations met for the first time in London in March 1998 at a European Conference, a new venture championed by Tony Blair's Labour government. Age-old controversies overshadowed even this most expansive, pan-European of events. Turkey, whose European credentials have been debated endlessly over centuries and which first applied to join the EC in 1987, was positioned below all of the other 11 countries in the EU's unofficial 'order of preference' and consequently boycotted the event. Facilitating the entry of these 11 countries will present enormous challenges. It is difficult to imagine how enlargement and deepening can both proceed simultaneously.

Summary

The process of economic integration in Europe since 1945 has been remarkably successful, despite the many tensions it has generated. A complex range of European institutions now exist which facilitate and enable an integrated economic system to operate in the western half of the continent. This process, which was initially both facilitated and limited by the prerogatives of Cold War geopolitics, rapidly developed its own, independent momentum. Since the collapse of communism, the various structures which make up the new European architecture are open to people from the former eastern bloc. New extensions to these structures will soon be constructed to accommodate these arrivals. However, and notwithstanding the 'Eurosceptic' fears of a federal superstate, the EU has still not developed beyond a relatively narrow economic agenda and has also singularly failed to capture the imagination of the European peoples. This was clear in the late 1940s when, despite hopes that the masses would spontaneously reject the old nation-states in favour of a European alternative, the great majority remained loyal to the existing structures. It is equally true today, as the growing doubts about the Maastricht Treaty (arguably the first European accord seriously to compromise national political sovereignty) and the failure of the EU in Yugoslavia make clear. Speculation about the 'end of the nation-state', even the 'end of geography', seems distinctly premature.[151] As Anthony Smith reminds us, every European nation has its national holidays, its shrines, ceremonials and monuments which mark out the cultural landscape of nationhood. There is, as yet, no such symbolic landscape for Europe as a whole; no European equivalent of Bastille Day, Guy Fawke's Night, or Remembrance Sunday. Revealingly, the only monuments which have begun to serve such a transcendent, European function are those which recall that most tragic and sombre of all European episodes: the Holocaust.[152]

Notes to Chapter 4

1 Milward (1993, 185); see, however, Laquer (1992). Accounts of the actual process of European integration are, of course, legion. See, as useful summaries from different perspectives and historical vantage points, Urwin (1989, 1995); Vaughan (1979) and Stirk (1996).
2 Gaddis (1987); Loth (1988), McCauley (1995) and Yergin (1978) are useful starting points. Ó Tuathail's short essay in Ó Tuathail *et al.* (1998, 47–57) is an excellent geographical introduction.
3 See, for illustrations of this approach, D. Campbell (1992); Ó Tuathail (1996).
4 The term is borrowed from J.J. Anderson (1992).
5 Nolte (1987).
6 Hobsbawm (1994, 144–77, especially 176).
7 McCauley (1995, 114).
8 McCauley (1995, 115–16).
9 McCauley (1995, 118–19). Churchill claimed to feel a pang of guilt at this: 'Might it not be thought rather cynical', he suggested to Stalin, 'if it seemed we had disposed of these issues, so fateful to millions of people, in such an off-hand manner?'.
10 Gaddis (1972; 1987); Loth (1988); McCauley (1995); Mastny (1979).
11 McCauley (1995, 121–2). On the Russo–Finnish border, see Paasi (1995).
12 Blacksell (1977, 25–6).
13 Jones (1994, 1–12).
14 Elkins (1988).
15 Hobsbawm (1994, 51); Jones (1994, 5).
16 Stirk (1996, 72–4).
17 H.B. Ryan (1982, 2); see also Streit (1939; 1941).

18 Quoted in P.J. Taylor (1990, 27); on Bevin, see Bullock (1983).
19 Deighton (1987); Greenwood (1986; 1992).
20 P.J. Taylor (1990, 51). Bevin later promoted the older, pan-regional idea of 'Euro–Africa', though subsumed within a bipolar system; see Kent (1989).
21 Lundestad (1986); see also Loth (1988); P.J. Taylor (1990, 44–55); R. Woods, (1990).
22 Churchill's relatively complex geopolitical ideal of a wider European federation acting alongside the British Empire within a pan-regional world order always jostled uncomfortably alongside his simpler, bipolar ideological view. In 1931, he spoke of 'the two great opposing forces of the future . . . [the] English-speaking peoples and Communism'; quoted in P.J. Taylor (1990, 28).
23 Quoted in Churchill (1974, vol. 7, 6761); McCauley (1995, 132); see also Harbutt (1986). The image of an 'iron curtain' had previously been used by Joseph Goebbels, the Nazi propaganda chief, to describe a Europe ripped in two by the alien, non-European powers of the USA and USSR which Nazi Germany was pledged to resist; see G. Parker (1988, 105).
24 McCauley (1995, 133–4).
25 Quoted in McCauley (1995, 136).
26 Roosevelt consistently rejected the idea that the USA should help to shape a post-war Europe, partly to preserve the wartime alliance with the USSR but also because of a genuine antipathy to imperialist interference in the peacetime affairs of other countries, a twentieth-century continuation of George Washington's insistence on US avoidance of 'entangling alliances'. See Harper (1996, 7–131) on Roosevelt, and Ó Tuathail and Agnew (1992); Ó Tuathail *et al.* (1998, 61–5) and Harper (1996, 135–232) on Kennan.
27 The United Nations was established as a more powerful alternative to the old League of Nations following discussions at Yalta and at an international conference at Dumbarton Oaks in Washington, DC, in 1945.
28 Quoted in Amen (1979, 5); McCauley (1995, 138–9); Ó Tuathail *et al.* (1998, 58–60); see also Alexander (1982).
29 Hobsbawm (1994, 48); P.J. Taylor (1990, 46).
30 Ambrose (1990).
31 Arkes (1972); Gimbel (1976); Hogan (1987); Mee (1984).
32 Quoted in McCauley (1995, 139).
33 Ellwood (1991); Woods and Jones (1991).
34 Stirk (1996, 90); Waever (1993, 154).
35 Quoted in McCauley (1995, 140–1).
36 Cromwell (1982).
37 Blacksell (1977, 37).
38 d'Attore (1991); Stirk (1991).
39 Stirk (1996, 87).
40 Milward (1984); Stirk (1996, 83–118).
41 Stirk (1996, 92 and 98–101).
42 Spain was allowed limited participation from 1954 and full membership after 1959, despite Franco's fascist regime. Yugoslavia, a socialist federation under the fiercely independent Josip Broz Tito, was involved as an observer.
43 Archer (1990, 33–41); Blacksell (1977, 47–9).
44 Quoted partially in Stirk (1996, 107).
45 Heller and Gillingham (1992); see also Archer (1990, 143–67); Blacksell (1977, 49–55).
46 Rostow (1960).
47 Haas (1958); Lindberg (1963).
48 Blacksell (1977, 57–81).
49 Childs (1983); A. Jones (1994, 13–33).

50 G. Smith (1995).
51 Stirk (1996, 101).
52 Kennedy (1988, 495).
53 Kennedy (1988, 510).
54 Bunge (1988, 133).
55 Barnaby (1985, 169); Kennedy (1988, 651).
56 Burgess (1991).
57 Lipgens and Loth (1991, 8–111). On the national movements, within and out-
    side the European Union of Federalists, see Lipgens and Loth (1988, 17–268,
    441–565 and 628–762); see also Burgess (1989).
58 Lipgens and Loth (1991, 277–318); see also Newman (1983).
59 Lipgens and Loth (1991, 186–276).
60 Lipgens and Loth (1991, 112–85).
61 Lipgens and Loth (1991, 319–435).
62 Lipgens (1982); Lipgens and Loth (1988, 662–6); Nelson and Stubb (1994,
    5–9). Churchill's position remained broadly consistent. Immediately after his re-
    election as Prime Minister in 1951 he informed his cabinet that: 'I am not
    opposed to a European Federation ... provided that it comes about naturally
    and gradually. But I never thought that Britain or the British Commonwealth
    should ... become an integral part of a European federation, and have never
    given the slightest support to the idea.' Two years later, he scribbled his vision of
    the most desirable world order on the back of a menu for the benefit of the
    German Chancellor, Konrad Adenauer. This showed three interlocking circles, a
    simplification of the Olympic motif, representing the USA, Britain and the
    Commonwealth, and the United States of Europe; see Greenwood (1992, 40–1);
    J.W. Young (1985a); Garton Ash (1993, 390).
63 Greece, Iceland, Turkey, the Federal Republic of Germany, Austria, Cyprus,
    Switzerland and Malta all joined over the next 14 years.
64 Quoted in Blacksell (1977, 85).
65 Archer (1990, 42–52).
66 Waever (1990).
67 Quoted in Stirk (1996, 122).
68 See McMillan (1992, 153–62) on the Fourth Republic.
69 Nelson and Stubb (1994, 11–12); see also Gillingham (1991).
70 Quoted in Gillingham (1991, ix); Stirk (1996, 121).
71 Lynch (1993).
72 Gillingham (1991, 148–77); see also Archer (1990, 53–62).
73 Greenwood (1992, 30–41); Ranieri (1991); J.W. Young (1985a). The United
    Kingdom subsequently joined as an associate member in December 1954.
74 Blacksell (1977, 132–44); A.M. Williams (1991, 14–29).
75 The ECSC comprised a Consultative Committee (representing producers, work-
    ers and consumers), a Council of Ministers (made up of ministerial representa-
    tives from each member state, the conduits between the ECSC and national
    governments), a Common Assembly (comprising elected parliamentarians from
    the national governments) and a Court of Justice (which included senior judges
    from each country, who ruled on the legal questions raised by the operation of
    the Paris treaty); see Gillingham (1991).
76 Blacksell (1977, 90–2); Stirk (1996, 129–33).
77 Stirk (1996, 136–46).
78 Archer (1990, 63–120); Blacksell (1977, 92–103).
79 Nelson and Stubb (1994, 13–15).
80 Greenwood (1992, 61–78).
81 Archer (1990, 121–32); Blacksell (1977, 107–31).
82   Quoted in Stirk (1996, 140); see also Beloff (1996, 51–83).

83 Milward (1993); see also Bulmer (1983); Groom (1978).
84 Milward (1992).
85 Milward and Sørensen (1993, 4).
86 Milward (1993, 198).
87 A.M. Williams (1991, 30–78) gives an exceptionally clear geographical reading of this period; see also P. Taylor (1983).
88 Stirk (1996, 169).
89 This argument was fully developed some years later by Servan-Schreiber (1967).
90 Stirk (1996, 163–6).
91 Nelson and Stubb (1994, 39–40).
92 Balfour (1992, 166–76).
93 The new EC inherited the same institutional architecture as its parent bodies: a Commission (made up of nominated representatives from the member states, each responsible for a particular aspect of European integration); a Council of Ministers (comprising national ministers from each member state who were to maintain communication between governments and the EC); a European Court of Justice (which adjudicated on legal questions and developed European law); and a European Assembly (unelected at this point but which was intended to develop into a directly elected parliament).
94 McMillan (1992, 177–84).
95 Wallerstein (1991, 65–83, 54).
96 For an interesting comment on violence and war in the shaping of ethnic identities, see A.D. Smith (1981).
97 Rokkan and Urwin (1982); C.H. Williams (1982; 1989). The campaign for Scottish and Welsh devolution has only recently achieved a measure of success following the election of a Labour government in May 1997 and two subsequent referendums in which a substantial majority of the Scottish population and a narrow majority of the Welsh people voted in favour of devolved national parliaments.
98 Garton Ash (1993) provides a brilliant, and strangely moving, account.
99 Balfour (1992, 180).
100 Blacksell (1977, 96–9).
101 Knox and Agnew (1994, 393–6).
102 George (1990, 66–9); see also Beloff (1996, 85–111); Greenwood (1992, 103–18).
103 J. Palmer (1987); M. Smith (1984).
104 Stirk (1996, 203–43); A.M. Williams (1991, 79–111).
105 Dalby (1988; 1990).
106 Dinan (1994, 109–19); Greenwood (1992, 108–112).
107 Nelson and Stubb (1994, 45–50).
108 Quoted in Harper (1996, 335). On Huntington, see Ó Tuathail (1996, 240–9).
109 Stirk (1996, 205).
110 Waever (1990). See, for an interesting commentary, Morin (1991).
111 Quoted in Stirk (1996, 207).
112 Nelson and Stubb (1994, 51–64). Subsidiarity has scarcely captured the popular imagination, however, not least because it remains a vague idea. There is no consensus, for example, over which decisions should be taken at what level. The sceptical British view was summed up by Nigel Lawson, sometime Chancellor of the Exchequer in the Thatcher cabinet: 'the Brussels Commission wants . . . to destroy the nation state – both ways. Not just through taking far more powers to the centre, to Brussels. But also by devolving powers, and they decide which powers it will be, to the lowest levels'; quoted in Stirk (1996, 231–2); see also Scott *et al.* (1994).

113 Betz (1990); Habermas (1992).
114 Kundera (1984); see also Croan (1989); R.J.W. Evans (1992); Garton Ash (1990); Gellner (1990); Hobsbawm (1991); Judt (1990); Le Rider (1994); Rupnik (1990); Schöpflin and Wood (1989); Stirk (1994a); Szücs (1988).
115 Seton-Watson (1985, 14); Davies (1996, 19–31).
116 Garton Ash (1993, 384).
117 Garton Ash (1993, 389).
118 Quoted in Stirk (1996, 249–50).
119 On the background to Gorbachev's European policy, see Neumann (1996, 131–93).
120 Hauner (1992, 10–11).
121 Chilton and Ilyin (1993); Malcolm (1989).
122 Pinder (1991).
123 Pinder (1991, 211).
124 Moravcsik (1991).
125 Milward and Sørensen (1993, 32).
126 Sandholtz and Zysman (1989).
127 This phrase is stolen from Michael Ondaatje's *The English Patient* (1992, 296).
128 Rössler (1994).
129 Ó Tuathail *et al.* (1998, 131–5 and, more generally, 103–78); see also C.H. Williams (1993).
130 O'Loughlin and van der Wusten (1990).
131 Fukuyama (1992). For an entirely different use of the term 'history' in this context, see Glenny (1990).
132 Duignan and Gann (1994, 316–18 and ix). Less celebratory, but still optimistic, readings can be found in Crouch and Marquand (1992).
133 See, for an intriguing discussion, Kumar (1992), but also see Brzezinski (1989–90) and Zeman (1991).
134 Halliday (1990).
135 Thernborn (1995, 299–300).
136 Hobsbawm (1992a, 165; 1994, 31); see also Hobsbawm (1992b).
137 Pajic (1993).
138 Mearsheimer (1990).
139 The mammoth scale of the economic task involved in the transition from communism to capitalism is outlined in Turnock (1997).
140 Balibar (1991); Ford (1992); Miles (1994); Wieviorka (1994).
141 Pearson (1995, 73).
142 Sidaway (1994).
143 These terms are stolen from Waever (1993, 207); see also Mlinar (1992).
144 Mestrovic (1993, 1994).
145 See D. Campbell (1997), together with Glenny (1993), Ó Tuathail (1996, 187–223), Popovski (1995) and Silber and Little (1996); see also Agnew (1994) and, for alternatives, Coakley (1994); Ruggie (1993) and P.J. Taylor (1994; 1995).
146 Ironically, it was the Serb community (both in Serbia itself and in Bosnia) that initially claimed to be fighting to preserve the federation through the Yugoslav army which was predominantly Serbian. The desire to create a 'Greater Serbia' soon asserted itself, however, and to date most of the worst atrocities, including those carried out in 1998 in the Serbian province of Kosovo, which has a predominantly Albanian population, have been perpetrated by Serbian forces.
147 Hungary joined the Council of Europe as early as November 1990.
148 See, as examples, Connolly (1995); Spicer (1992).

149 Pinder (1991); Van Ham (1993).
150 Five of these – Hungary, the Czech Republic, Poland, Slovenia and Estonia –
     began formal negotiations to join the EU in July 1997 and seem likely to join in
     the first 'wave'.
151 O'Brien (1992); Ohmae (1990; 1995).
152 A.D. Smith (1992); see also Cerutti (1992); Hodgson (1993); J. Young (1993).

Conclusion: the idea of Europe in the twenty-first century

> Ambition, covetousness, irresolution, fear and desires do not abandon us just because we have changed our landscape. They often follow us into the very cloister and schools of philosophy. . . . [I]t is not enough to withdraw from the mob, not enough to go to another place: we have to withdraw from such attributes of the mob as are within us. It is our own self we have to isolate and take back into possession.
>
> Michel de Montaigne, *On Solitude* (1580)

This book has recounted a tale of dramatic, and often violent transformations in Europe's political geography. The maps included here are simply snapshots which hint at the kaleidoscopic changes that have taken place. The accompanying text has sought to describe the different ideas about Europe which have informed these changes. This too has been a saga of upheaval and conflict, part of the wider ebb and flow of what Parker once called 'tidal Europe'.[1] But we have also encountered some remarkably persistent geopolitical instincts that have informed and limited the development of the European idea through the ages. Chief amongst them has been a tenacious territoriality, an enduring European conviction that human existence can only be fulfilling and meaningful if it is rooted in a particular place, a specific and bounded geographical area (whether it call itself a city, a region, a province or a nation) which has bestowed upon it, usually as a result of violent struggle, the mystique of a 'homeland'. Despite repeated predictions that these 'primeval' instincts will collapse under the influence of geography-defying technologies (from the 'modern' continental railways, telegraphs and aeroplanes of the 1890s to the 'postmodern' computers, satellites and mobile phones of the 1990s) the territorial imperative has retained much of its potency. The periodic attempts to create a transcendent European unity, whether by force or through consensus, have always confronted this powerful geopolitical tradition, an inertia which has resisted tyrants and visionaries alike.

The post-1945 attempt to build an economic union in western Europe has been more successful and benign than earlier attempts and has certainly modified the 'frontier of national sovereignty' in the western half of the continent.[2] But this took place within the strictly limited geopolitical circumstances of the Cold War and within a divided Europe. Its success has also been constrained by the mainly economic nature of this integration, though it is precisely this emphasis which has allowed the process to proceed thus far with so little democratic accountability. Arguably, the entire project would have foundered had it been subjected to regular electoral scrutiny. The 'democratic deficit' of the European programme still remains. How many people can name their European Member of Parliament? Recent attempts to stray beyond the economic agenda have required greater popular consultation and this has demonstrated the persistence of national particularism. The EU, like its precursors, is nothing more than the sum of the constituent nation-states which created it, largely for traditional nationalist reasons. It does not, nor can it, exist independently of these states. It therefore shares many (if not all) of their collective failings. As John Lambert has persuasively argued, the EU is a complex interstate bureaucracy which, like the state bureaucracies which produced it, tends to shut off genuinely radical alternative perspectives in the endless search for compromise.[3] The same powerful lobbies and vested interests dominate economic and social policy-making at both the European and the national level. The groups denied a voice in Brussels are precisely those who are disenfranchised by governments in Paris, Berlin, London, Rome or Madrid.

We cannot assume, therefore, that an integrated or even a united Europe will automatically produce a more tolerant, inclusive and cosmopolitan society. The central dilemma, I would suggest, lies in the fact that the old nation-states and the emerging EU rest on common territorial assumptions. These have been implicit in many European negotiations about immigration and citizenship rights which have reflected rather than challenged national fears that 'outsiders' will undermine the purity and integrity of the 'homeland'. Lurking behind the rhetoric of unity and peace, the idea of Europe has always been about the politics of exclusion and division; the drawing of arbitrary lines between those within and those without.[4] This tendency was all too obvious during the recent Yugoslavian crisis and has now become central to the entire European debate. The removal of the Cold War ideological division has reopened the ancient geographical question about the eastern limits of Europe, an inquiry which goes far deeper than a simple exercise in cartographic delimitation. The presence of the Soviet Union and its empire was the single most important factor promoting economic and political co-operation in western Europe after 1945, and the growing doubts about the pace and direction of the European project since 1991, though certainly linked to the Maastricht Treaty and all that it entails, are also connected to a basic geopolitical question: what purpose does Europe serve in the absence of a perceived Soviet threat? To put it more generally: can Europe exist without an obvious constituting 'Other'?

It is important to emphasise the possibilities for alternative forms of European politics operating beyond the formal structures of the EU and the various nation-states, the kind of popular, campaigning social movements, rooted in civil society, which have recently developed in relation to nuclear disarmament, the environment and the rights of minority groups.[5] Perhaps the most important dimension of such movements is their potential to work within an 'anti-geopolitical' framework, one which does not draw, nor depend, on the territoriality which has limited the European debate thus far. Some observers have detected the beginnings of a deterritorialised political activity arising from the new regional dynamics of the European economy. Christopher Harvie, for example, has written of a 'Europe of the regions', a flexible, interactive system in which city regions in different parts of Europe develop common programmes and agendas within the EU (or beyond it) independently of the old nation-states in which they are located. In this way, traditional territoriality is 'unbundled' and the idea of the state as a 'container', in which one is either 'in' or 'out', is subverted. 'We are dealing', claims Harvie, 'not so much with a revolt against the central state, as an attempt to dismember it and rearticulate it in new ways'.[6] The regions which are beginning to function in this way, however, are all in the wealthier and most dynamic parts of Europe – Lombardy, Rhône–Alpes, Baden–Württemberg – where a new breed of entrepreneur–politician has produced what Harvie calls 'bourgeois regionalism'. Hesitant though they are, these inchoate regional systems have led one authority to suggest that 'the economic geography of the late 20th century is coming to resemble the economic geography of the Middle Ages', a return, in a very different guise, to the Medieval city regions and the older trading alliances such as existed under the Hanseatic League.[7]

This theme extends arguments made in the late 1970s by Hedley Bull about the possibility of a deterritorialised political consciousness which, though thoroughly (post)modern, would reflect the fluid and overlapping spatialities of the distant past, what is sometimes called 'the new Medievalism'.[8] The implication is that European citizens should more readily develop complex, multiple and international allegiances within and beyond their place of residence, linkages which may or may not have a geographical dimension.[9] This in turn implies an urgent need to embrace, indeed to celebrate, those circumstances which we have been taught to regard only in negative terms. The ideas of exile or diaspora, for example, generally suggest deviations from desired norms of 'rootedness' and belonging. So long as that attitude persists, the prospects of building a truly multicultural and cosmopolitan European society are bleak indeed. Such a society requires, by definition, a rejection of the tyranny of place and an acceptance that cherished and apparently timeless ethnicities and spatialities are in fact mere inventions. This would produce what Appadurani calls a 'postnational geography', a diasporic Europe of overlapping, changing and temporary identities, an arena of endless heterogeneity and hybridity, a place to which no one could lay exclusive claim and from which no one could be automatically excluded, in short a truly cosmopolitan society of the kind described so movingly by Yi Fu Tuan:

> Thinking ... yields a twofold gain: although it isolates us from our immediate group it can link us both seriously and playfully with the cosmos – to strangers in other places and times; and enable us to accept a human condition that we have always been tempted by fear and anxiety to deny, namely, the impermanence of our state wherever we are, our ultimate homelessness. A cosmopolite is one who considers the gain greater than the loss. Having seen something of the splendid spaces, he or she ... will not want to return, permanently, to the ambiguous safeness of hearth.[10]

The idea that the answer to Europe's problems lies in re-creating patterns of flexible allegiance from the ancient past is, of course, more than a little fanciful.[11] But the argument that we need to 'un-think' Europe, to set both the term and the idea free from the intellectual moorings which have tied them so firmly to older territorial structures, remains a powerful one.[12] In the end, the appropriate words were written by Jurgen Habermas:

> our task is less to reassure ourselves of our common origins in the European Middle Ages than to develop a new political self-confidence commensurate with the role of Europe in the world of the twenty-first century. Hitherto, world history has accorded the empires that have come and gone only *one* appearance on the stage. ... It now appears as if Europe as a whole is being given a second chance. It will not be able to make use of this in terms of the power politics of yesteryear, but only under changed premises, namely a non-imperial process of reaching understanding with, and learning from, other cultures.[13]

Notes to Conclusion

1 W.H. Parker (1968, 27–9).
2 The term in quotation marks is borrowed from Milward *et al.* (1993), though the authors disagree with the claim made here; see Milward (1993).
3 Lambert (1994).
4 Brubaker (1990); Husbands (1988); Meehan (1993); Silverman (1992); de Wenden (1997).
5 See the essays by Simon Dalby and Paul Routledge in Ó Tuathail *et al.* (1998, 179–87 and 245–55), together with Miall (1993; 1994).
6 Harvie (1994, 5); see also J.J. Anderson (1992); J. Anderson and Goodman (1995); P.J. Taylor (1991; 1994; 1995).

7 Marquand (1994, 21).
8 See Chapter 1 and Bull (1977). For recent discussions, see J. Anderson (1996); Brownlee *et al.* (1991) and Camilleri and Falk (1992).
9 Halliday (1988). Hoskyns (1996) gives an intriguing analysis of gender and European integration.
10 Tuan (1996, 188); see also Appadurani (1996) and, as examples of the expanding literature on these themes, Brah (1996); Comaroff (1991); Gilroy (1993); Hall (1990; 1992) and Pogge (1992).
11 It is worth noting that another kind of 'Medievalism' has recently been imagined by Samuel Huntington (1996). According to his prediction, Europe's return to a distant past would spell not a radical reworking of its internal geopolitics but rather the re-establishment of old struggles between a white, Christian Europe and the threatening non-white, non-Christian areas beyond, in North Africa and the Middle East. This would represent a 'clash of civilisations' between an aging, rich population protected by a 'wall of prosperity' from a young, poor one. This is a disturbing vision, all the more disquieting if the xenophobia which this suggests were ever to be unleashed against Europe's minority peoples.
12 Nederveen Pieterse (1991; 1994); Pocock (1991).
13 Habermas (1992, 12–13).

Bibliography

Abraham, D. (1981) *The Collapse of the Weimar Republic: Political Economy and Crisis* (Princeton).

Adamthwaite, A.P. (1980) *The Lost Peace: International Relations in Europe 1918–39* (London).

Adas, M. (1989) *Machines as the Measure of Man: Science, Technology, and Ideologies of Western Dominance* (Ithaca).

Ageron, C.-R. (1975) 'L'idée d'Eurafrique et le débat colonial Franco–Allemand de l'entre-deux-guerres', *Revue d'Histoire Moderne et Contemporaine* 22: 446–75.

Agnelli, G. and Cabiata, A. (1919) [1918] *Fédération européenne ou Ligue des Nations?* (Paris).

Agnew, J. (1989) 'The Devaluation of Place in Social Science', in J. Agnew and J.S. Duncan (eds), *The Power of Place: Bringing Together the Geographical and Sociological Imaginations* (London), 9–29.

Agnew, J. (1994) 'The Territorial Trap: The Geographical Assumptions of International Relations Theory', *Review of International Political Economy* 1, 1: 53–80.

Agnew, J. and Corbridge, S. (1995) *Mastering Space: Hegemony, Territory and International Political Economy* (London).

Akerman, J.R. (1995) 'The Structuring of Political Terrain in Early Printed Atlases', *Imago Mundi* 47: 138–55.

Albrecht-Carrié, R. (1965) *The Unity of Europe: An Historical Survey* (London).

Albrecht-Carrié, R. (ed.) (1968) *The Concert of Europe, 1815–1914* (London).

Aldcroft, D.H. (1997) *Studies in the Interwar European Economy* (Aldershot).

Alexander, G.M. (1982) *The Prelude to the Truman Doctrine: British Policy in Greece, 1944–47* (Oxford).

Ambrose, S.E. (1990) *Rise to Globalism: American Foreign Policy since 1938* (Harmondsworth).

Amen, M.P. (1979) *American Foreign Policy in Greece 1944–49* (London).

Amin, S. (1989) *Eurocentrism* (London).

Ancel, J. (1936) *Géopolitique* (Paris).

Ancel, J. (1936–45) *Manuel géographique de la politique européenne.* Vol. 1 (1936) *L'Europe centrale*; Vol. 2, Part 1 (1940) *L'Europe germanique et ses bornes*; Vol. 2, Part 2 (1945) *L'Allemagne* (Paris).

Ancel, J. (1938) *Géographie des frontières* (Paris).

Anderson, B. (1991) [1983] *Imagined Communities: Reflections on the Origin and Spread of Nationalism* (London).

Anderson, J. (1996) 'The Shifting Stage of Politics: New Medieval and Postmodern Territorialities?', *Environment and Planning D: Society and Space* 14: 133–53.

Anderson, J. and Goodman, J. (1995) 'Regions, States and the European Union: Modernist Reaction or Postmodern Adaptation', *International Review of Political Economy* 2, 4: 600–31.

Anderson, J.J. (1992) *The Territorial Imperative: Pluralism, Corporatism and Economic Crisis* (Cambridge).

Anderson, M.S. (1954) 'English Views of Russia in the Seventeenth Century', *The Slavonic and East European Review* 39: 143–53.

Anderson, M.S. (1970) 'Eighteenth Century Theories of the Balance of Power', in R. Hatton and M.S. Anderson (eds), *Studies in Diplomatic History: Essays in Memory of David Bayne Horn* (London), 183–98.

Anderson, M.S. (1976) *Europe in the Eighteenth Century 1713–83* (2nd edn (London).

Andrew, C. (1985) *Secret Service: The Making of the British Intelligence Service* (London).

Andrew, C. and Kanya-Forstner, A.J. (1981) *France Overseas: The Great War and the Climax of French Imperial Expansion* (London).

Anon. (1919) 'Proposed New Administrative Subdivisions of England, France and Germany', *Geographical Review* 7: 114–18.

Appadurani, A. (1996) 'Sovereignty Without Territoriality: Notes for a Postnational Geography', in P. Yaeger (ed.), *The Geography of Identity* (Ann Arbor), 40–58.

Archer, C. (1990) *Organizing Western Europe* (London).

Arkes, H. (1972) *Bureaucracy, the Marshall Plan and National Interest* (Princeton).

Atkinson, D. (1995) 'Geopolitics, Cartography and Geographical Knowledge: Envisioning Africa from Fascist Italy', in M. Bell, R.A. Butlin and M. Heffernan (eds), *Geography and Imperialism, 1820–1940* (Manchester), 265–97.

Atkinson, D. (1996) Geopolitics and the Geographical Imagination in Fascist Italy, unpublished PhD thesis, University of Loughborough, Loughborough.

Aulneau, J. (1926) *Histoire de l'Europe centrale* (Paris).

Baczko, B. (1984) 'Le calendrier républicain: décréter l'éternité', in P. Nora (ed.), *Les lieux de mémoire*. Vol. 1: *La République* (Paris), 37–83.

Baechler, J., Hall, J.A. and Mann, M. (eds) (1988) *Europe and the Rise of Capitalism* (Oxford).

Balfour, M. (1992) *Germany: The Tides of Power* (London).

Balibar, E. (1991) 'Es gibt Keinan Staat in Europa: Racism and Politics in Europe Today', *New Left Review* 186: 5–19.

Balzaretti, R. (1992) 'The Creation of Europe', *History Workshop Journal*, 33: 181–96.

Banks, M. (1996) *Ethnicity: Anthropological Constructions* (London).

Bann, S. (1984) *The Clothing of Clio: A Study of the Representation of History in Nineteenth-Century Britain and France* (Cambridge).

Banton, M. (1987) *Racial Theories* (Cambridge).

Barbiche, B. (1978) *Sully* (Paris).

Barker, E. (1914) *Nietzsche and Treitschkte: The Worship of Power in Modern Germany* (Oxford).

Barker, E. (1915) *The Submerged Nationalities of the German Empire* (Oxford).

Barker, E. (1918) *Linguistic Oppression in the German Empire* (London).

Barnaby, F. (1985) 'Nuclear Free Zones', in D. Pepper and A. Jenkins (eds), *The Geography of Peace and War* (Oxford).

Barraclough, G. (1963) *European Unity in Thought and Practice* (Oxford).

Barraclough, G. (ed.) (1984) *The Times Atlas of World History* (London).

Bartlett, R. (1993) *The Making of Europe: Conquest, Colonization, and Cultural Change 950–1350* (London).

Bartlett, R. and Mackay, A. (1989) *Medieval Frontier Societies* (Oxford).

Bassin, M. (1987a), 'Imperialism and the Nation State in Friedrich Ratzel's Political Geography', *Progress in Human Geography* 11: 473–95.

Bassin, M. (1987b), 'Race contra Space: The Conflict Between German Geopolitik and National Socialism', *Political Geography Quarterly* 6: 115–34.

Bassin, M. (1988) 'Expansion and Colonialism on the Eastern Frontier: Views of Siberia and the Far East in pre-Petrine Russia', *Journal of Historical Geography* 14: 3–21.

Bassin, M. (1991) 'Russia Between Europe and Asia: The Ideological Construction of Geographical Space', *Slavonic and East European Review* 50: 1–17.

Bassin, M. (1992) 'Geographical Determinism in fin-de-siècle Marxism: Georgii Plekhanov and the Environmental Basis of Russian History', *Annals of the Association of American Geographers* 82: 3–22.

Bassin, M. (1993) 'Turner, Solov'ev, and the "Frontier Thesis": the Nationalist Significance of Open Spaces', *Journal of Modern History* 65: 473–511.

Bassin, M. (1994) 'Russian Geographers and the "National Mission" in the

Far East', in D. Hooson (ed.), *Geography and National Identity* (Oxford), 112–33.

Bauman, Z. (1985) 'On the Origins of Civilisation: A Historical Note', *Theory, Culture and Society* 2: 7–14.

Bauman, Z. (1989) *Modernity and the Holocaust* (Ithaca).

Bayly, C. (ed.) (1989) *Atlas of the British Empire: A New Perspective on the British Empire from 1500 to the Present* (London).

Bellers, J. (1710) *Some Reasons for an European State* (London).

Beloff, M. (1957) *Europe and the Europeans* (London).

Beloff, M. (1967) *The Balance of Power* (Montreal).

Beloff, M. (1996) *Britain and the European Union* (London).

Beneton, P. (1975) *Histoire des mots: culture et civilisation* (Paris).

Berdiaev, N.A. (1924) *The New Middle Ages* (London).

Berdiaev, N.A. (1946) *The Russian Idea* (London).

Berg, P. and Lodgaard, S. (1983) 'Disengagement Zones: A Step Towards Meaningful Defence', *Journal of Peace Research* 20: 5–15.

Bergson, H. (1915) *La signification de la guerre* (Paris).

Berki, R.N. (1989) 'Marxism and European Unity', in P.M.R. Stirk (ed.), *European Unity in Context: The Interwar Period* (London), 41–64.

Bessel, R. (1984) *Political Violence and the Rise of Nazism: The Stormtroopers in Eastern Germany, 1925–1934* (London).

Bessel, R. (ed.) (1987) *Life in the Third Reich* (Oxford).

Betz, H.E. (1990) 'Mitteleuropa and Post-Modern European Identity', *New German Critique* 50: 173–92.

Beveridge, W.M. (1940) *Peace by Federation* (London).

Bhabha, H. (ed.) (1990) *Nation and Narration* (London).

Biddis, M.D. (1970) *Father of Racist Ideology: The Social and Political Thought of Count Gobineau* (London).

Bishop, P. (1989) *The Myth of Shangri-La: Tibet, Travel Writing and the Western Creation of a Sacred Landscape* (London).

Black, J. (1994) *Convergence or Divergence? Britain and the Continent* (London).

Blacksell, M. (1977) *Post-War Europe: A Political Geography* (London).

Blaut, J.M. (1993) *The Colonizer's Model of the World: Geographical Diffusion and Eurocentric History* (New York).

Blouet, B.W. (1976) 'Sir Halford Mackinder as British High Commissioner in South Russia, 1919–20', *Geographical Journal* 142: 228–46.

Blouet, B.W. (1987) *Sir Halford Mackinder: A Biography* (College Station, TX).

Blouet, B.W. (1996) 'The Political Geography of Europe: 1900–2000 A.D.', *Journal of Geography* 95, 1: 5–14.

Bluntschli, J.K. (1871) *Europa als Staatenbund* (Bonn).

Bourke, J. (1996) *Dismembering the Male: Men's Bodies, Britain and the Great War* (London).

Bowman, I. (1921) *The New World: Problems of Political Geography* (New York).

Bowman, I. (1942) 'Geography vs. Geopolitics', *Geographical Review* 32: 646–58.

Boyce, R. (1980) 'Britain's First "No" to Europe', *European Studies Review* 10: 17–45.

Boyce, R. (1989) 'British Capitalism and the Idea of European Unity between the Wars', in P.M.R. Stirk (ed.), *European Unity in Context: The Interwar Period* (London), 65–83.

Brah, A. (1996) *Cartographies of Diaspora: Contesting Identities* (London).

Brailsford, H.N. (1914) *The War of Steel and Gold* (London).

Brailsford, H.N. (1939) *The Federal Idea* (London).

Branford, V. and Geddes, P. (1919a) [1917] *The Coming Polity: A Study in Reconstruction* (London).

Branford, V. and Geddes, P. (1919b), *Our Social Inheritance* (London).

Braudel, F. (1972) [1949], *The Mediterranean and the Mediterranean World in the Age of Philip II* (London).

Braudel, F. (1979) *Civilization and Capitalism, 15th–18th Century* 3 vols, (London).

Braudel, F. (1986) *L'Identité de la France: espace et histoire* (Paris).

Brauer, R.W. (1995) 'Boundaries and Frontiers in Medieval Muslim Geography', *Transactions of the American Philosophical Society*, 85, 6: 1–73.

Braun, F. and Ziegfeld, A.H. (1929–30) *Geopolitischer Geschichtsatlas* (Dresden).

Brechtefeld, J. (1996) *Mitteleuropa in German Politics, 1848 to the Present* (London).

Breuilly, J. (1992) 'The National Idea in Modern German History', in J. Breuilly (ed.), *The State in Germany: The National Idea in the Making, Unmaking and Remaking of a Modern Nation-State* (London), 1–28.

Bridenthal, R., Grossman, A. and Kaplan, M. (eds) (1984) *When Biology Became Destiny: Women in Weimar and Nazi Germany* (New York).

Brigham, A.P. (1919) 'Principles in the Determination of Boundaries', *Geographical Review* 7, 4: 201–19.

Brinken, A.D. (1973) 'Europa in der Kartographie des Mittelalters', *Archiv für Kulturdeschichte*, 55: 289–304.

Broers, M. (1996) *Europe Under Napoleon 1799–1815* (London).

Broszat, M. (1981) *The Hitler State* (London).

Broszat, M. (1987) *Hitler and the Collapse of Weimar Germany* (Leamington Spa).

Brownlee, M.S., Brownlee, K. and Nichols, S.G. (eds) (1991) *The New Medievalism* (Baltimore).

Brubaker, W.R. (1990) 'Immigration, Citizenship and the Nation-State in France and Germany: A Comparative Historical Analysis', *International Journal of Sociology* 5, 4: 379–407.

Brunhes, J. and Vallaux, C. (1921) *La géographie de l'histoire: géographie de la paix et de la guerre sur terre et sur mer* (Paris).

Bryce, J. (1922) *Modern Democracies* (London).

Brzezinski, Z. (1989–90) 'Post-Communist Nationalism', *Foreign Affairs* 68, 5: 1–12.

Bugge, P. (1993) 'The Nation Supreme: The Idea of Europe 1914–1945', in K. Wilson and J. van der Dussen (eds), *The History of the Idea of Europe* (London), 83–149.

Buisseret, D. (1968) *Sully and the Growth of Centralized Government in France 1598–1610* (London).

Buisseret, D. (1984) *Henry IV, King of France* (London).

Buisseret, D. (ed.) (1992) *Monarchs, Ministers and Maps: The Emergence of Cartography as a Tool of Government in Early Modern Europe* (Chicago).

Bull, H. (1977) *The Anarchical Society: A Study of Order in World Politics* (London).

Bull, H., Kingsbury, B. and Roberts, A. (eds.) (1990) *Hugo Grotius and International Relations* (Oxford).

Bullock, A. (1962) [1952] *Hitler: A Study in Tyranny* (Harmondsworth).

Bullock, A. (1983) *Ernest Bevin: Foreign Secretary 1945–1951* (Oxford).

Bullock, A. (1991) *Hitler and Stalin: Parallel Lives* (London).

Bulmer, S. (1983) 'Domestic Politics and European Community Policy Making', *Journal of Common Market Studies* 21: 349–63.

Bunge, W. (1988) *Nuclear War Atlas* (Oxford).

Burgess, M. (1989) *Federalism and European Union* (London).

Burgess, M. (1991) 'Innocents Abroad: Federalism and the Resistance between Party and Movement 1945–47', P.M.R. Stirk and D. Willis (eds), *Shaping Postwar Europe: European Unity and Disunity 1945–1957* (London), 7–14.

Burke, E. (1983) [1790] *Reflections on the Revolution in France and on the Proceedings in Certain Societies in London relative to that Event* (Harmondsworth).

Burke, P. (1980) 'Did Europe Exist before 1700?', *History of European Ideas* 1: 21–9.

Burke, P. (1992) 'We, the People: Popular Culture and Popular Identity in Modern Europe', in S. Lash and J. Friedman (eds), *Modernity and Identity* (Oxford), 293–308.

Burleigh, M. (1988) *Germany Turns Eastwards: A Study of Ostforschung in the Third Reich* (Cambridge).

Burleigh, M. (1994) *Death and Deliverance: 'Euthanasia' in Germany 1900–1945* (Cambridge).

Burleigh, M. and Wipperman, W. (1991) *The Racial State: Germany 1933–1945* (Cambridge).

Butler, N.M. (1932) *The International Mind: An Argument for the Judicial Settlement of International Disputes* (New York).

Butler, N.M. (1940) *Why War? Essays and Addresses on War and Peace* (New York).

Cahnman, W. (1952) 'Frontiers between East and West', *Geographical Review* 49: 605–24.

Camilleri, J. and Falk, J. (1992) *The End of Sovereignty? The Politics of a Shrinking and Fragmenting World* (Aldershot).

Campbell, D. (1992) *Writing Security: The United States Foreign Policy and the Politics of Identity* (Minneapolis).

Campbell, D. (1997) *National Deconstruction: Violence, Identity and Justice in Bosnia* (Minneapolis).

Campbell, H. (1918) *The Biological Aspects of Warfare* (London).

Cannadine, D. (1981) 'War and Death, Grief and Mourning in Modern Britain', in J. Whaley (ed.), *Mirrors of Mortality: Studies in the Social History of Death* (London), 187–242.

Carr, E.H. (1981) [1939] *The Twenty Years' Crisis, 1919–1939: An Introduction to the Study of International Relations* (London).

Ceadel, M. (1987) *Pacifism in Britain, 1914–45: The Defining of a Faith* (Oxford).

Cerutti, F. (1992) 'Can there be a Supranational Identity?', *Philosophy and Social Criticism*, 18, 2: 147–62.

Chabod, F. (1965) *Storia dell'idea d'Europa* (Bari).

Chadwick, H.M. (1945) *The Nationalities of Europe and the Growth of National Ideologies* (Cambridge).

Chamberlain, H.S. (1915) *Who is to Blame for the War?* (New York).

Chamberlin, J.E. and Gilman, S.L. (eds) (1985) *Degeneration: The Dark Side of Progress* (New York).

Charles-Brun, J. (1911) *Le régionalisme* (Paris).

Charlesworth, A. (1994) 'Contesting Places of Memory: The Case of Auschwitz', *Environment and Planning D: Society and Space* 12, 51: 579–94.

Childers, T. (1983) *The Nazi Voter: The Social Formations of Fascism in Germany, 1919–33* (Chapel Hill).

Childs, D. (1988) [1983] *The GDR: Moscow's German Ally* (London).

Chilton, P. and Ilyin, M. (1993) 'Metaphor in Political Discourse: The Case of "Common European House"', *Discourse and Society* 4: 7–31.

Chiriot, D. (ed.) (1989) *The Origins of Backwardness in Eastern Europe: Economics and Politics from the Middle Ages until the Early Twentieth Century* (Berkeley and Los Angeles).

Chisholm, G.G. (1917a) 'Central Europe: A Review', *Scottish Geographical Magazine* 33: 83–8.

Chisholm, G.G. (1917b) 'Central Europe as an Economic Unit', *Geographical Teacher* 9: 122–33.

Chisholm, G.G. (1919) 'The Geographical Prerequisites of a League of Nations: A Review', *Scottish Geographical Magazine* 35: 248–56.

Churchill, W.S. (1974) *Winston S. Churchill: His Complete Speeches, 1897–1963* (ed. R. Rhodes James) (London).

Clark, G. (ed.) (1987) *John Bellers, 1654 to 1725: Quaker Visionary – His Life, Times and Writings* (York).

Clarke, C.H. (ed.) (1916) *The Ravings of a Renegade, Being the War Essays of H.S. Chamberlain* (London).

Clarke, D., Doel, M. and Mcdonough, F.X. (1996) 'Holocaust Topologies: Singularity, Politics, Space', *Political Geography* 15: 457–89.

Claval, P. (1994) 'Playing with Mirrors: The British Empire according to Albert Demangeon', in A. Godlewska and N. Smith (eds), *Geography and Empire* (Oxford), 228–43.

Clout, H.D. (1996) *After The Ruins: Restoring the Countryside of Northern France after the Great War* (Exeter).

Coakley, J. (1994) 'Approaches to the Resolution of Ethnic Conflicts: The Strategy of Non-Territorial Autonomy', *International Political Science Review*, 15: 297–314.

Cobb, R. (1987) *The People's Armies. The armées révolutionnaires: instrument of the Terror in the departments, April 1793 to Floréal Year II* (New Haven).

Cofrancesco, D. (1985) 'Ideas of the [Italian] Fascist Government and Party on Europe', in W. Lipgens (ed.), *Documents on the History of European Integration* (Berlin) Vol. 1, 177–84.

Cohn, N. (1996) [1967] *Warrant for Genocide: The Myth of the Jewish World Conspiracy and the Protocols of the Elders of Zion* (London).

Cole, G.D.H. (1947) *Local and Regional Government* (London).

Cole, T. and Smith, G. (1995) 'Ghettoization and the Holocaust: Budapest 1944', *Journal of Historical Geography* 21, 3: 300–16.

Colley, L. (1992) *Britons: Forging the Nation 1707–1837* (New Haven).

Comaroff, J. (1991) 'Humanity, Ethnicity, Nationality: Conceptual and Comparative Perspectives on the USSR', *Theory and Society*, 20: 661–88.

Comité d'Études (1918–19) *Travaux du Comité d'Études*. Vol. 1: *L'Alsace–Lorraine et la frontière du Nord-Est*. Vol. 2: *Questions Européennes – Belgique, Slesvig, Tchécoslovaquie, Pologne et Russie; Questions Adriatiques – Yougoslavie, Roumanie, Turquie d'Europe et d'Asie* (Paris).

Conley, T. (1996) *The Self-Made Map: Cartographic Writing in Early Modern France* (Minneapolis).

Connolly, B. (1995) *The Rotten Heart of Europe* (London).

Conquest, R. (1986) *The Harvest of Sorrow* (London).

Cormack, L. (1991) 'Good Fences Make Good Neighbours: Geography as Self-Definition in Early-Modern England', *Isis* 82: 639–61.

Corni, G. (1990) *Hitler and the Peasants: Agrarian Policy of the Third Reich 1930–1939* (Oxford).

Cornish, V. (1936) *Borderlands of Language in Europe and their Relation to the Historic Frontier of Christendom* (London).

Coudenhove-Kalergi, R.N. (1923) *Pan-Europa* (Vienna).

Coudenhove-Kalergi, R.N. (1943) *Crusade for Pan-Europe* (New York).

Coudenhove-Kalergi, R.N. (1953) *An Idea Conquers the World* (London).

Craig, J.W. (1984) *Scholarship and Nation-Building: The Universities of Strasbourg and Alsatian Society, 1870–1939* (Chicago).

Croan, M. (1989) 'Lands In-Between: the Politics of Cultural Identity in Contemporary Eastern Europe', *Eastern European Politics and Society* 3: 176–97.

Cromwell, W.C. (1982) 'The Marshall Plan, Britain and the Cold War', *Review of International Studies* 8: 238–42.

Cronin, V. (1989) *Paris on the Eve 1900–1914* (London).

Crook, P. (1994) *Darwinism, War and History: The Debate Over the Biology of War from the 'Origin of Species' to the First World War* (Cambridge).

Crouch, C. and Marquand, D. (eds) (1992) *Towards Greater Europe? A Continent Without an Iron Curtain* (Oxford).

Curry, W.B. (1939) *The Case for a Federal Europe* (Harmondsworth).

Curtis, L. (1950) [1935] *Civitas Dei* (London).

Curzon, G.N. (1899) *Russia in Central Asia in 1889* (London).

Curzon, G.N. (1907) *Frontiers* (Oxford).

Cutler, A.C. (1991) 'The "Grotian Tradition" in International Relations', *Review of International Studies* 17: 41–65.

Cvijic, J. (1916) *Questions balkaniques* (Paris).

Cvijic, J. (1918) *La péninsule balkanique: géographie humaine* (Paris).

Dalby, S. (1988) 'Geopolitical Discourse: the Soviet Union as Other', *Alternatives* 13: 415–42.

Dalby, S. (1990) *Creating the Second Cold War: The Discourse of Politics* (London).

Dalby, S. (1991) 'Critical Geopolitics: Discourse, Difference, and Dissent', *Environment and Planning D: Society and Space* 93: 261–83.

Daniels, S. (1993) *Fields of Vision: Landscape Imagery and National Identity in England and the United States* (Cambridge).

Darwin, L. (1926) *The Need for Eugenic Reform* (London).

d'Attore, P.P. (1991) 'Americanism and Anti-Americanism in Italy', in P.M.R. Stirk and D. Willis (eds), *Shaping Postwar Europe: European Unity and Disunity 1945–1957* (London), 43–52.

Davies, N. (1996) *Europe: A History* (London).

de Grazier, V. (1992) *How Fascism Ruled Women: Italy, 1922–1945* (Berkeley and Los Angeles).

d'Eichthal, G. (1840) *De l'unité européenne* (Paris).

Deighton, A. (1987) 'The "Frozen Front": The Labour Government, the Division of Germany and the Origins of the Cold War 1945-7', *International Affairs* 63: 449–65.

de Jouvenel, B. (1930) *Vers les États Unis d'Europe* (Paris).

Delaisi, F. (1929) *Les deux Europes* (Paris).

Delanty, G. (1995) *Inventing Europe: Idea, Identity, Reality* (London).

de Lapradelle, P. (1928) *La frontière: étude de droit international* (Paris).

de Launay, L. (1917) *France–Allemagne* (Paris).

Demangeon, A. (1920) *Le déclin de l'Europe* (Paris).

Demangeon, A. (1923) *L'Empire britannique: étude de géographie coloniale* (Paris).

Demangeon, A. (1931) 'Les bases économiques d'une entente européenne', *Annales d'Histoire Économique et Sociale* 11: 449–54.

Demangeon, A. (1932) 'Les conditions géographiques d'une Union européenne: fédération européenne ou ententes régionales', *Annales d'Histoire Économique et Sociale* 17: 433–51.

Demangeon, A. and Febvre, L. (1935) *Le Rhin: problèmes d'histoire et d'économie* (Paris)

de Martonne, E. (1930) *L'Europe centrale* 2 vols (Paris).

der Boer, P. (1993) 'Europe to 1914: The Making of an Idea', in K. Wilson and J. van der Dussen (eds), *The History of the Idea of Europe* (London), 13–82.

Derolez, A. (ed.) (1968) *Liber floridus* (Ghent).

de Saint-Pierre, C.I.C. (1712) *Mémoires pour rendre la paix perpétuelle en Europe* (Cologne).

de Saint-Pierre, C.I.C. (1728) *A Discourse of the Danger of Governing by One Minister* (London).

de Saint-Simon, C.H. and Thierry, A. (1814) *De la reorganisation de la société européenne* (Paris).

Devine, A. (1921) *Off The Map: The Suppression of Montenegro – The Tragedy of a Small Nation* (London).

de Wenden, C. (1997) *La citoyenneté européenne* (Paris).

Dickenson, R.E. (1943) *The German Lebensraum* (London).

Dickenson, R.E. (1947) *City, Region and Regionalism* (London).

Diehl, J.M. (1987) 'Victors or Victims? Disabled Veterans in the Third Reich', *Journal of Modern History* 59: 705–36.

Digeon, C. (1959) *La crise allemande de la pensée française (1870–1914)* (Paris).

Dijkink, G. (1996) *National Identity and Geopolitical Visions: Maps of Pride and Pain* (London).

Dilke, O.A.W., Millard, A.R. and Aujar, G. (1987) 'Cartography in Ancient Europe and the Mediterranean', in J.B. Harley and D. Woodward (eds), *The History of Cartography*. Vol. 1 *Cartography in Prehistoric, Ancient and Medieval Europe and the Mediterranean* (Chicago), 103–280.

Dinan, D. (1994) *Ever Closer Union? An Introduction to the European Community* (London).

Dion, R. (1947) *Les frontières de la France* (Paris).

Dion, R. (1977) *Aspects politiques de la géographie antique* (Paris).

Dionisotti, C. (1971) *Europe in Sixteenth-Century Italian Literature* (Oxford).

Dockès, P. (1969) *L'Espace dans la pensée économique du XVIe. au XVIIIe. siècle* (Paris).

Dodgshon, R.A. (1987) *The European Past: Social Evolution and Spatial Order* (London).

Dominian, L. (1917) *Frontiers of Language and Nationality in Europe* (New York).

Dorpalen, A. (1942) *The World of General Haushofer: Geopolitics in Action* (New York).

Drouet, J. (1912) *L'Abbé de Saint-Pierre: l'homme et l'oeuvre* (Paris).

Droz, J. (1960) *L'Europe centrale: évolution historique de l'idée de 'Mitteleuropa'* (Paris).

Dryer, C.R. (1920) 'Mackinder's "World Island" and its American '"Satellite"', *Geographical Review* 9: 205–7.

Duhamel, G. (1931) [1930] *America the Menace* (London).

Duignan, P. and Gann, L.H. (1994) *The United States and the New Europe 1945–1993* (Oxford).

Dunbar, G. (1974) 'Elisée Reclus and the Great Globe', *Scottish Geographical Magazine* 90: 57–66.

Duroselle, J.-B. (1966) *L'Idée d'Europe dans l'histoire* (Paris).

Duroselle, J.-B. (1990) *Europe: A History of its Peoples* (London).

Eatwell, R. (1996) *Fascism: A History* (London).

Edgerton, S.Y. (1987) 'From Mental Matrix to Mappamundi to Christian Empire: The Heritage of Ptolemaic Cartography in the Renaissance', in D. Woodward (ed.), *Art and Cartography: Six Historical Essays* (Chicago), 10–50.

Eksteins, M. (1989) *Rites of Spring: The Great War and the Birth of the Modern Age* (London).

Eley, G. (1988) 'Nazism, Politics and the Image of the Past', *Past and Present* 121: 171–208.

Elias, N. (1994) *The Civilizing Process: The History of Manners and State Formation and Civilization* (Oxford).

Elkins, T.H. (1988) *Berlin: The Spatial Structure of a Divided City* (London).

Elliot, J.H. (1992) 'Europe of Composite Monarchies', *Past and Present*, 137: 48–71.

Ellwood, D. (1991) 'The Marshall Plan and the Politics of Growth', in P.M.R. Stirk and D. Willis (eds), *Shaping Postwar Europe: European Unity and Disunity 1945–1957* (London), 15–26.

Evans, A. (1916) 'The Adriatic Slavs and the Overland Route to Constantinople', *Geographical Journal* 47, 4: 241–65.

Evans, R.J.W. (1992) 'Frontiers and National Identities in Central Europe', *International History Review* 14, 3: 480–502.

Fahlbusch, M., Rössler, M. and Siegrist, D. (1989) 'Conservatism, Ideology and Geography in Germany 1920–1950', *Political Geography Quarterly* 8: 353–67.

Fawcett, C.B. (1917) 'Natural Divisions of England', *Geographical Journal* 49, 2: 124–41.

Fawcett, C.B. (1918) *Frontiers: A Study in Political Geography* (Oxford).

Fawcett, C.B. (1919) *Provinces of England: A Study of Some Geographical Aspects of Devolution* (London).

Fawcett, C.B. (1933) *Political Geography of the British Empire* (London).

Fawcett, C.B. (1940) 'Some Geographical Factors in World Unity', *New Commonwealth Quarterly* 6: 95–101.

Fawcett, C.B. (1941) *The Basis of a World Commonwealth* (London).

Febvre, L. (1974) 'Civilisation: Evolution of a Word and a Group of Ideas', in P. Burke (ed.), *A New Kind of History from the Writings of Febvre* (New York), 219–57.

Figes, O. (1996) *A People's Tragedy: The Russian Revolution 1891–1924* (London).

Fink, C. (1989) *Marc Bloch: A Life in History* (Cambridge).

Fischer, F. (1967) *Germany's Aims in the First World War* (London).

Fischer, J. (1957) *Oriens–Occidens–Europa: Begriff und Gedanke 'Europa' in der Späteren Antike und im frühen Mittelaltar* (Wiesbaden).

Fisher, A.W. (1970) *The Russian Annexation of the Crimea, 1772–1783* (Cambridge).

Fisher, C.A. (1966) 'The Changing Dimensions of Europe', *Journal of Contemporary History*, 1, 3: 3–20.

Fisher, H.A.L. (1927) *Life of James Bryce, Viscount Bryce of Dechmont* (London).

Fisher, H.A.L. (1936) *A History of Europe* (London).

Fitzgerald, F.S. (1986) [1934], *Tender is the Night* (London).

Fletcher, R. (1984) *Revisionism and Empire: Socialist Imperialism in Germany, 1897–1914* (London).

Fleure, H.J. (1915) *Regional Surveys in Relation to Geography* (Oxford).

Fleure, H.J. (1916) 'France: A Regional Interpretation', *Scottish Geographical Magazine* 32: 519–35.

Fleure, H.J. (1916–17) 'Berlin and its Region', *Sociological Review* 9: 14–26.

Fleure, H.J. (1917) 'Régions humaines', *Annales de Géographie* 26: 161–74.

Fleure, H.J. (1918) *Human Geography of Western Europe: A Study in Appreciation* (London).

Fleure, H.J. (1921) *The Treaty Settlement of Europe: Some Geographic and Ethnographic Aspects* (Oxford).

Flory, T. (1966) *Le mouvement régionaliste français: sources et développements* (Paris).

Foerster, R.M. (1967) *Europa: Geschichte einer politischen Idee* (Munich).

Ford, G. (1992) *Fascist Europe: The Rise of Racism and Xenophobia* (London).

Foster, R.F. (1989) *Modern Ireland 1600–1972* (Harmondsworth).

Foucault, M. (1991) 'Governmentality', in G. Burchell, C. Gordon and P. Miller (eds), *The Foucault Effect: Studies in Governmentality* (Chicago), 87–104.

Fox, E.W. (1992) *The Emergence of the Modern European World* (Oxford).

France, P. (1985) 'Western Civilization and its Mountain Frontiers (1750–1850)', *History of European Ideas* 6, 3: 297–310.

Freeman, M. (1995) [1987] *Atlas of Nazi Germany: A Political, Economic and Social Anatomy of the Third Reich* 2nd edn (London).

Frei, N. (1993) *Nazi Germany: A Social History* (Oxford).

French, P. (1995) *Younghusband* (London).

Fry, A.R. (1935) *John Bellers 1654–1725: Quaker, Economist and Social Reformer* (London).

Fuhrmann, M. (1981) *Europa – zur Geschichte einer kulturellen und politischen Idee* (Constance).

Fukuyama, F. (1992) *The End of History and the Last Man* (Harmondsworth).

Fulbrook, M. (ed.) (1994) *National Histories and European Histories* (London).

Fussell, P. (1975) *The Great War and Modern Memory* (Oxford).

Gaddis, J.L. (1972) *The United States and the Origins of the Cold War, 1941–1947* (New York).

Gaddis, J.L. (1987) *The Long Peace: Inquiries into the History of the Cold War* (Oxford).

Galbo, J. (1996) 'Sex, Geography, and Death: Metropolis and Empire in a Fascist Writer', *Environment and Planning D: Society and Space* 14: 35–58.

Gallois, L. (1918) 'Alsace–Lorraine and Europe', *Geographical Review* 6: 89–115.

García, S. (ed.) (1993) *European Identity and the Search for Legitimacy* (London)

Garton Ash, T. (1990) 'Mitteleuropa', *Daedalus* 119, 1: 1–22.

Garton Ash, T. (1993) *In Europe's Name: Germany and the Divided Continent* (London).

Geddes, P. (1915) 'Wardom and Peacedom: Suggestions towards an Interpretation', *Sociological Review* 8: 275–93.

Geddes, P. [aka Cites Committee, Sociological Society] (1918), 'A Rustic View of War and Peace', *Sociological Review* 10, 1: 1–24.

Geddes, P. and Branford, V. (1916–17) 'The Making of the Future', *Sociological Review* 9: 100–4.

Geddes, P. and Slater, G. (1917) *Ideas at War* (London).

Gelfand, L.E. (1963) *The Inquiry: American Preparations for Peace, 1917–1919* (New Haven).

Gellner, E. (1990) 'Ethnicity and Faith in Eastern Europe', *Deadalus* 119, 1: 279–94.

George, S. (1990) *An Awkward Partner: Britain and the European Community* (Oxford).

Geremek, B. (1996) *The Common Roots of Europe* (Cambridge).

Gilbert, E.W. (1939) 'Practical Regionalism in England and Wales', *Geographical Journal* 94: 29–44.

Gilbert, E.W. (1952) 'Geography and Regionalism', in G. Taylor (ed.), *Geography in the Twentieth Century: A Study of Growth, Fields, Techniques, Aims and Trends* (London), 345–71.

Gilbert, F. (1951) 'The "New Diplomacy" of the Eighteenth Century', *World Politics* 4: 1–38.

Gilbert, M. (1972) *Imperial Russian History Atlas* (London).

Gilbert, M. (1978) *The Holocaust: Maps and Photographs* (New York).

Gilbert, M. (1982) *Atlas of the Holocaust* (London).

Gilbert, M. (1986) *The Holocaust: The Jewish Tragedy* (London).

Gilbert, M. (1989) *Second World War* (London).

Gilbert, M. (1994) *First World War* (London).

Gill, A. (1993) *A Dance Between Flames: Berlin Between the Wars* (London).

Gillingham, J. (1985) *Industry and Politics in the Third Reich: Ruhr Coal, Hitler and Europe* (London).

Gillingham, J. (1991) *Coal, Steel, and the Rebirth of Europe, 1945–1955: The Germans and French from Ruhr Conflict to Economic Community* (Cambridge).

Gillis, J.R. (ed.) (1994) *Commemorations: The Politics of National Identity* (Princeton).

Gilroy, P. (1993) *The Black Atlantic: Modernity and Double Consciousness* (London).

Gimbel, J. (1976) *The Origins of the Marshall Plan* (Stanford).

Glacken, C. (1967) *Traces on the Rhodian Shore: Nature and Culture in Western Thought from Ancient Times to the End of the Eighteenth Century* (Berkeley and Los Angeles).

Glenny, M. (1990) *The Rebirth of History: Eastern Europe in the Age of Democracy* (Harmondsworth).

Glenny, M. (1993) *The Fall of Yugoslavia: The Third Balkan War* (Harmondsworth).

Goblet, Y. (1936) *The Twilight of Treaties* (London).

Godlewska, A. and Smith, N. (eds) (1994) *Geography and Empire* (Oxford).

Goldhagen, D.J. (1996) *Hitler's Willing Executioners: Ordinary Germans and the Holocaust* (London).

Gollwitzer, H. (1951) *Europabild und Europagedanke* (Munich).

Grant, M. (1916) *The Passing of the Great Race; Or, The Racial Basis of European History* (New York).

Gravier, G. (1919) *Les frontières historiques de la Serbie* (Paris).

Gregory, A. (1994) *The Silence of Memory: Armistice Day 1919–1946* (Oxford).

Gregory, D. (1994) *Geographical Imaginations* (Oxford).

Greenhalgh, P. (1988) *Ephemeral Vistas: The Expositions Universelles, Great Exhibitions and World's Fairs, 1851–1939* (Manchester).

Greenwood, S. (1986) 'Bevin, the Ruhr and the Division of Germany, August 1945–December 1946', *Historical Journal* 29: 203–12.

Greenwood, S. (1992) *Britain and European Cooperation since 1945* (Oxford).

Grenville, J.A.S. (1986) [1976] *Europe Reshaped 1848–1878* (London).

Griffin, R. (1991) *The Nature of Fascism* (London).

Groom, A.J.R. (1978) 'Neofunctionalism: A Case of Mistaken Identity', *Political Science* 30: 15–28.

Grosrichard, A. (1979) *Structure du sérail: la fiction du despotisme asiatique dans l'Occident classique* (Paris).

Gross, F. (1948) 'The Peace of Westphalia, 1648–1948', *American Journal of International Law* 42: 20–41.

Gruffudd, P. (1994) 'Back to the Land: Historiography, Rurality and the Nation in Interwar Wales', *Transactions of the Institute of British Geographers New Series* 19: 61–77.

Guenée, B. (1986) 'Des limites féodales aux frontières politiques', in P. Nora (ed.), *Les lieux de mémoire*. Vol. 2, Part 2. *La nation* (Paris), 11–33.

Gyorgy, A. (1944) *Geopolitics: The New German Science* (Berkeley and Los Angeles).

Haas, E.B. (1958) *The Uniting of Europe: Political, Social and Economic Forces, 1950–1957* (Stanford).

Habermas, J. (1992) 'Citizenship and National Identity: Some Reflections on the Future of Europe', *Praxis International* 12, 1: 1–19.

Hale, J. (1993) 'The Renaissance Idea of Europe', in S. Garcia (ed.), *European Identity and the Search for Legitimacy* (London), 46–63.

Halecki, O. (1950) *The Limits and Divisions of European History* (London).

Halecki, O. (1952) *Borderlands of Western Civilization: A History of East Central Europe* (New York).

Hall, P. (1996) [1988], *Cities of Tommorrow: An Intellectual History of Urban Planning and Design in the Twentieth Century* (Oxford).

Hall, S. (1990) 'Cultural Identity and Diaspora', in J. Rutherford (ed.), *Identity: Community, Culture, Difference* (London), 222–37.

Hall, S. (1992) 'New Ethnicities', in J. Donald and A. Rattansi (eds), *'Race', Culture and Difference* (London), 252–9.

Halliday, F. (1988) 'Three Concepts of Internationalism', *International Affairs* 64: 187–98.

Halliday, F. (1990) 'The Ends of the Cold War', *New Left Review* 180: 5–23.

Harbutt, F. (1986) *The Iron Curtain: Churchill, America and the Origins of the Cold War* (Oxford).

Harley, J.B. (1988) 'Secrecy and Silences: The Hidden Agenda of Cartography in Early Modern Europe', *Imago Mundi*, 40: 111–30.

Harley, J.B. and Woodward, D. (eds) (1992) *The History of Cartography*. Vol. 2, Part 1. *Cartography in the Traditional Islamic and South Asian Societies* (Chicago).

Harley, J.B. and Woodward, D. (eds) (1994) *The History of Cartography*. Vol. 2, Part 2. *Cartography in the Traditional East and Southeast Asian Societies* (Chicago).

Harper, J.L. (1996) *American Visions of Europe: Franklin D. Roosevelt, George F. Kennan and Dean G. Acheson* (Cambridge).

Hartley, J.M. (1992) 'Is Russia Part of Europe? Russian Perceptions of Europe in the Reign of Alexander I', *Cahiers du Monde Russe et Sovietique* 33: 369–86.

Harvey, P.D.A., Woodward, D., Campbell, T. and Harley, J.B. (1987) 'Cartography in Medieval Europe and the Mediterranean', in J.B. Harley and D. Woodward (eds), *The History of Cartography*. Vol. 1 *Cartography in Prehistoric, Ancient, and Medieval Europe and the Mediterranean* (Chicago), 283–509.

Harvie, C. (1991) 'English Regionalism: The Dog that Never Barked', in B. Crick (ed.), *National Identities* (Oxford), 105–19.

Harvie, C. (1994) *The Rise of Regional Europe* (London).

Hassinger, H. (1917) 'Das geographische Wesen Mitteleuropas, nebst einigen grundsätzlichen Bemerkungen über die geographischen Naturgebiete Europas und ihre Begrenzung', *Mitteilungen der Geographischen Gesellschaft Wien* 60: 437–93.

Hauner, M. (1992) [1990] *What is Asia to Us? Russia's Asian Heartland Yesterday and Today* (London).

Haushofer, K. (1924) *Geopolitik des Pazifischen Ozeans* (Berlin).

Haushofer, K. (1927) *Grenzen* (Berlin).

Haushofer, K. (1930) 'Mitteleuropa und der Anschluss', in F. Kleinwaechter and H. Paller (eds), *Der Anschluss* (Vienna), 147–59.

Haushofer, K. (1930–34) *Macht und Erde*. Vol. 1 (1930) *Die Grossmächte vor und nach dem Weltkriege*; Vol. 2 (1932) *Jenseits der Grossmächte*; Vol. 3 (1934) *Raumüberwindende Mächte* (Leipzig and Berlin).

Haushofer, K. (1937) 'Mitteleuropa und die Welt', *Zeitschrift für Geopolitik* 14: 1–4.

Haushofer, K. (1941) *Der Kontinentalblock: Mitteleuropa – Eurasien – Japan* (Munich).

Hay, D. (1968) [1958] *Europe: The Emergence of an Idea* 2nd edn (Edinburgh).

Hay, D. (1980) 'Europe Revisited: 1979', *History of European Ideas* 1: 1–6.

Hayward, J.E.S. (1973) *The One and Indivisible French Republic* (London).

Hazard, P. (1990) *The European Mind: the Critical Years 1680–1715* (New York).

Heater, D. (1992) *The Idea of European Unity* (Leicester).

Hechter, M. and Brustein, W. (1980) 'Regional Modes of Production and Patterns of State Formation in Western Europe', *American Journal of Sociology* 85: 1061–94.

Heffernan, M. (1995) 'For Ever England: The Western Front and the Politics of Remembrance in Britain', *Ecumene* 2, 3: 293–323.

Heffernan, M. (1996) 'Geography, Cartography and Military Intelligence: The Royal Geographical Society and the First World War', *Transactions of the Institute of British Geographers* New Series 21: 504–33.

Heller, F.H. and Gillingham, J.R. (eds) (1992) *NATO: The Founding of the Atlantic Alliance and the Integration of Europe* (New York).

Henderson, W.O. (1983) *Friedrich List: Economist and Visionary 1789–1846* (London).

Henig, R. (1995) [1985] *Versailles and After 1919–1933* (London).

Hepple, L. (1986) 'The Revival of Geopolitics', *Political Geography Quarterly*, Supplement to 5, 4: 21–36.

Herb, G.H. (1997) *Under the Map of Germany: Nationalism and Propaganda 1918–1945* (London).

Herbertson, A.J. (1905) 'The Major Natural Regions of the World: An Essay in Systematic Geography', *Geographical Journal* 25: 200–12.

Herbertson, A.J. (1913) 'The Higher Units: A Geographical Essay', *Scientia* 14: 203–13.

Herbertson, A.J. (1913–14) 'Natural Regions', *Geographical Teacher* 7: 158–63.

Herf, J. (1984) *Reactionary Modernism: Technology, Culture and Politics in Weimar and the Third Reich* (Cambridge).

Herford, O. (1919) *This Giddy Globe* (New York).

Herriot, E. (1930) *The United States of Europe* (London).

Herrmann, D.G. (1996) *The Arming of Europe and the Making of the First World War* (Princeton).

Hertslet, E. (1875–91) *The Map of Europe by Treaty; Showing the Various Political and Territorial Changes Which Have Taken Place Since The General Peace of 1814* (4 Vols, London).

Heske, H. (1986) 'German Geographic Research in the Nazi Period', *Political Geography Quarterly* 5: 267–82.

Heske, H. (1987) 'Karl Haushofer: His Role in German Geopolitics and Nazi Politics', *Political Geography Quarterly* 6: 135–44.

Hettner, A. (1915a) *Englands Weltherrschaft und der Krieg* (Leipzig).

Hettner, A. (1915b) *Die Ziele unserer Weltpolitik (Der Deutsche Krieg* No. 64) (Leipzig).

Hettner, A. (1916) 'Das Britische und das Russische Reich' *Geographische Zeitschrift* 25: 353–71.

Hettner, A. (1917a), *Der Friede und die deutsche Zukunft* (Stuttgart).

Hettner, A. (1917b), 'Deutschlands territoriale Neugestaltung', *Geographische Zeitschrift* 25: 57–72.

Hinsley, F.H. (1966) 'The Concert of Europe', in Lawrence W. Martin (ed.), *Diplomacy in Modern European History* (London), 43–57.

Hinsley, F.H. (1967) *Power and the Pursuit of Peace* (Cambridge).

Hobhouse, L.J. (1916) *Questions of War and Peace* (London).

Hobsbawm, E.J. (1991) 'The Return of Mitteleuropa', *Guardian* 11 Oct.

Hobsbawm, E.J. (1992a) *Nations and Nationalism since 1780* (Cambridge).

Hobsbawm, E.J. (1992b) 'Ethnicity and Nationalism in Europe Today', *Anthropology Today* 8, 1: 3–8.

Hobsbawm, E.J. (1994) *Age of Extremes: The Short History of the Twentieth Century* (London).

Hobson, J.A. (1916) *Towards International Government* (London).

Hodgson, M.G.S. (1993) 'Grand Illusion: the Failure of European Consciousness', *World Policy Journal* 10, 2: 13–8.

Hogan, M.J. (1987) *The Marshall Plan* (London).

Holdar, S. (1992) 'The Ideal State and the Power of Geography: the Life and Work of Rudolf Kjellén', *Political Geography* 11: 307–23.

Holdich, T.H. (1909) 'Some Aspects of Political Geography', *Geographical Journal* 34: 593–607.

Holdich, T.H. (1916a), *Political Boundaries and Boundary Making* (London).

Holdich, T.H. (1916b), 'Geographical Problems in Boundary Making', *Geographical Journal* 47, 6: 421–40.

Holdich, T.H. (1916c), 'Political Boundaries', *Scottish Geographical Magazine* 32: 497–507.

Holdich, T.H. (1918a), *Boundaries in Europe and the Near East* (London).

Holdich, T.H. (1918b), 'Presidential Address', *Geographical Journal* 52, 1: 1–12.

Holdich, T.H. (1919) 'Presidential Address', *Geographical Journal* 54, 1: 1–12.

Hooson, D. (ed.) (1994) *Geography and National Identity* (Oxford).

Hopkirk, P. (1994) *On Secret Service East of Constantinople: The Plot to Bring Down the British Empire* (London).

Hosking, G. (1997) *Russia: People and Empire 1552–1917* (London).

Hoskyns, C. (1996) *Integrating Gender: Women, Law and Politics in the European Union* (London).

House, E.M. and Seymour, C. (eds) (1921) *What Really Happened at Paris: The Story of the Peace Conference, 1918–1919 by American Delegates* (New York).

Houston, J.M. (1953) *A Social Geography of Europe* (London).

Hume, D. (1994) [1752] 'The Idea of a Perfect Commonwealth', reprinted in G. Claeys (ed.), *Utopias of the British Enlightenment* (Cambridge), 55–69.

Hunter, J.M. (1983) *Perspectives on Ratzel's Political Geography* (Lanham, Maryland).

Huntington, E. (1907) *The Pulse of Asia: A Journey in Central Asia illustrating the Geographic Basis of History* (London).

Huntington, E. (1915) *Civilization and Climate* (London).

Huntington, E. (1919) *World Power and Evolution* (New Haven).

Huntington, E. (1924) *The Character of Races as Influenced by Physical*

Environment, Natural Selection, and Historical Development (New York).

Huntington, S.P. (1996) *The Clash of Civilizations and the Re-making of the World Order* (London).

Husbands, C. (1988) 'The Dynamics of Racial Exclusion and Expulsion: Racist Politics in Western Europe', *European Journal of Political Research* 16: 701–20.

Hynes, S. (1990) *A War Imagined: The First World War and English Culture* (London).

Inglis, K.S. (1993) 'Entombing Unknown Soldiers from London to Paris to Baghdad', *History and Memory* 5, 2: 7–31.

Ionescu, G. (ed.) (1976) *The Political Thought of Saint Simon* (Oxford).

Jacob, C. (1992) *L'Empire des cartes: approche théoretique de la cartographie à travers l'histoire* (Paris).

James, H. (1986) *The German Slump: Politics and Economics, 1924–1936* (Oxford).

Jennings, I.W. (1940) *A Federation of Western Europe* (London).

Jewsbury, G. (1976) *The Russian Annexation of Bessarabia, 1774–1783* (Boulder, CA).

Johnson, D.W. (1915) 'Geographic Aspects of the War', *Geographical Review* 12: 431–84.

Johnson, D.W. (1917a), 'The Role of Political Boundaries', *Geographical Review* 4: 208–13.

Johnson, D.W. (1917b), *The Peril of Prussianism* (New York).

Johnson, D.W. (1917c), *Lettre d'un Américain à un Allemand sur la guerre et les responsabilités de l'Allemagne* (Paris).

Johnson, D.W. (1918) *Topography and Strategy in War* (London).

Johnson, D.W. (1921) *Battlefields of the World War, Western and Southern Fronts: A Study in Military Geography* (New York).

Johnson, H.J. (1919) 'The Anthropologist's Approach: Race, Language and Nationality in Europe', *Sociological Review* 11: 37–46.

Joll, J. (1980) 'Europe – an Historian's View', *History of European Ideas*, 1: 7–19.

Jones, A. (1994) *The New Germany: A Human Geography* (Chichester).

Jones, E.L. (1987) [1981] *The European Miracle: Environments, Economies, and Geopolitics in the History of Europe and Asia* 2nd edn (Cambridge).

Jones, P. (1988) *The Peasantry in the French Revolution* (Cambridge).

Jordan, D.S. (1915) *War and the Breed* (Boston).

Josephy, F.L. (1944) *Europe: The Key to Peace* (London).

Judt, T. (1990) 'The Rediscovery of Central Europe', *Daedalus* 119, 1: 23–54.

Kain, R.J.P. and Baigent, E. (eds) (1992) *The Cadastral Map in the Service of the State: A History of Property Mapping* (Chicago).

Kann, R.A. (1964) *The Multinational Empire: Nationalism and National*

Reform in the Habsburg Monarchy 1848–1918. Vol. 2 *Empire Reform* (New York).

Karrow, R.W. Jr (1993) *Mapmakers of the Sixteenth Century and Their Maps: Bio-Bibliographies of the Cartographers of Abraham Ortelius, 1570* (Chicago).

Kater, M. (1983) *The Nazi Party: A Social Profile of Members and Leaders 1919–1945* (Oxford).

Kearns, G. (1984) 'Closed Space and Political Practice: Frederick Jackson Turner and Halford Mackinder', *Environment and Planning D: Society and Space* 21: 23–34.

Kearns, G. (1985) 'Halford John Mackinder 1861–1947', in T.W. Freeman (ed.), *Geographers: Bio-Bibliographical Studies* vol. 9 (London), 71–86.

Kearns, G. (1993) 'Fin de siècle geopolitics: Mackinder, Hobson and Theories of Global Closure', in P.J. Taylor (ed.), *Political Geography of the Twentieth Century: A Global Analysis* (London), 9–30.

Kearns, G. (1997) 'The Imperial Subject: Geography and Travel in the Work of Mary Kingsley and Halford Mackinder', *Transactions of the Institute of British Geographers* New Series 22: 450–72.

Kemp, A. (1981) *The Maginot Line: Myth and Reality* (London).

Kennedy, P. (1988) *The Rise and Fall of the Great Powers: Economic Change and Military Conflict from 1500 to 2000* (London).

Kent, J. (1989) 'Bevin's Imperialism and the Idea of Eur-Africa, 1945–9', in M. Dockrill and J.W. Young (eds), *British Foreign Policy 1945–56* (London), 47–76.

Kern, S. (1983) *The Culture of Time and Space 1880–1918* (London).

Kershaw, I. (1987) *The 'Hitler Myth': Image and Reality in the Third Reich* (Oxford).

Kershaw, I. (1993) [1985] *The Nazi Dictatorship: Problems and Perspectives of Interpretation* (London).

Keynes, J.M. (1920) *The Economic Consequences of the Peace* (New York).

Kiernan, V.G. (1980) 'Europe in the Colonial Mirror', *History of European Ideas* 1, 1: 39–61.

Kitchen, M. (1988) *Europe Between the Wars: A Political History* (London).

Kjellén, R. (1924) [1916], *Der Staat als Lebensform* (Berlin).

Knox, P. and Agnew, J. (1994) *The Geography of the World Economy* (London).

Koch, H.W. (ed.) (1984) *The Origins of the First World War: Great Power Rivalry and German War Aims* (London).

Kohn, H. (1960) *Panslavism: Its History and Ideology* (New York).

Konvitz, J.W. (1987) *Cartography in France, 1660–1848: Science, Engineering, and Statecraft* (Chicago).

Konvitz, J.W. (1990) 'The State, Paris and Cartography in Eighteenth- and Nineteenth-Century France', *Journal of Historical Geography* 16, 1: 3–16.

Korinman, M. (1990) *Quand l'Allemagne pensait le monde: grandeur et décadence d'une géopolitique* (Paris).

Kost, K. (1988) *Die Einflüsse der Geopolitik auf Forschung und Theorie der Politischen Geographie von ihren Anfängen bis 1945. Ein Beitrag zur Wissenschaftsgeschichte der Politischen Geographie und ihre Terminologie unter Berücksichtigung von Militär- und Kolonialgeographie* (Bonn).

Kost, K. (1989) 'The Conception of Politics in Political Geography and Geopolitics in Germany until 1945', *Political Geography Quarterly* 8: 369–85.

Kristof, L.D. (1959) 'The Nature of Frontiers and Boundaries', *Annals of the Association of American Geographers* 49: 269–82.

Kristof, L.D. (1960) 'The Origins and Evolution of Geopolitics', *Journal of Conflict Resolution* 4, 1: 15–51.

Kristof, L.D. (1968) 'The Russian Image of Europe', in C.A. Fisher (ed.), *Essays in Political Geography* (London), 345–87.

Kroeber, A.L. and Kluckhohn, C. (1952) *Culture: A Critical Review of Concepts and Definitions* (New York).

Krüger, P. (1989) 'European Ideology and European Reality: European Unity and German Foreign Policy in the 1920s', in P.M.R. Stirk (ed.), *European Unity in Context: The Interwar Period* (London), 84–98.

Kumar, K. (1992) 'The 1989 Revolutions and the Idea of Europe', *Political Studies* 40: 439–61.

Kundera, M. (1984) 'The Tragedy of Central Europe', *New York Review of Books* 26 April, 33–8.

Lambert, J. (1994) *Solidarity and Survival: A Vision for Europe* (Aldershot).

Lampropoulos, V. (1993) *The Rise of Eurocentrism: Anatomy of Interpretation* (Princeton).

Langhorne, R. (1981) *The Collapse of the Concert of Europe* (London).

Laqueur, W. (1992) *Europe in Our Time: A History 1945–1992* (Harmondsworth).

Laqueur, W. (1996) 'Fin de Siècle: Once More With Feeling', *Journal of Contemporary History* 31: 5–49.

Laski, H. (1932) *Nationalism and the Future of Civilization* (London).

Lebovics, H. (1992) *True France: The Wars over Cultural Identity, 1900–1945* (Ithaca).

Le Bras, H. (1986) 'La Statistique Générale de la France', in P. Nora (ed.), *Les lieux de mémoire*. Vol. 2, Part 2. *La nation* (Paris), 317–53.

Lederer, I.J. (1963) *Yugoslavia at the Paris Peace Conference: A Study in Frontiermaking* (New Haven).

Lee, S.J. (1984) *Aspects of European History 1494–1789* (London).

Lefebvre, H. (1991) *The Production of Space* (Oxford).

Lemonon, E. (1931) *La nouvelle Europe centrale et son bilan économique* (Paris).

Le Rider, J. (1994) *La Mitteleuropa* (Paris).

Lestringant, F. (1994) *Mapping the Renaissance World: The Geographical Imagination in the Age of Discovery* (Cambridge).

Levy, C. (1996) *Italian Regionalism: History, Identity and Politics* (Oxford).

Lévy, J. (1997) *Europe: Une Géographie* (Paris).

Levy, J.S. (1983) *War in the Modern Great Power System, 1495–1975* (Lexington, KY).

Leyser, K. (1992) 'Concepts of Europe in the Early and High Middle Ages', *Past and Present*, 137: 25–47.

L'Héritier, M. (1928) 'Régions historiques: Europe centrale, Orient méditerranéen et question d'Orient', *Revue de Synthèse Historique* 45: 43–67.

Lindberg, L.N. (1963) *The Political Dynamics of European Economic Integration* (Stanford).

Lipgens, W. (1982) *A History of European Integration 1945–7*. Vol. 1. *The Formation of the European Unity Movement* (Oxford).

Lipgens, W. (ed.) (1985) *Documents on the History of European Integration*. Vol. 1. *Continental Plans for European Union 1939–1945* (Berlin).

Lipgens, W. (ed.) (1986) *Documents on the History of European Integration*. Vol. 2. *Plans for European Union in Great Britain and in Exile 1939–1945* (Berlin).

Lipgens, W. and Loth, W. (eds) (1988) *Documents on the History of European Integration*. Vol. 3. *The Struggle for European Union by Political Parties and Pressure Groups in Western European Counties 1945–1950* (Berlin).

Lipgens, W. and Loth, W. (eds) (1991) *Documents on the History of European Integration*. Vol. 4: *Transnational Organizations of Political Parties and Pressure Groups in the Struggle for European Union, 1945–1950* (Berlin).

List, F. (1904) *National Systems of Political Economy* (London).

Livingstone, D.N. (1992) *The Geographical Tradition: Episodes in the History of a Contested Enterprise* (Oxford).

Livey, J. (1981) 'The Europe of the Enlightenment', *History of European Ideas* 1: 91–102.

Lodgaard, S. and Thee, M. (eds) (1984) *Nuclear Disengagement in Europe* (London).

Lorimer, J. (1884) *The Institutes of the Law of Nations* (Edinburgh).

Loth, W. (1988) *The Division of the World 1941–55* (London).

Lothian, Lord [P. Kerr] (1935) *The Ending of Armageddon* (London).

Louis, W.R. (1967) *Great Britain and Germany's Lost Colonies 1914–1919* (Oxford).

Lundestad, G. (1986) 'Empire by Invitation? The United States and Western Europe, 1945–1952', *Journal of Peace Research* 23: 263–77.

Lyde, L.W. (1913) *The Continent of Europe* (London).

Lyde, L.W. (1914) 'Some Rough War Notes', *Geographical Journal* 44, 4: 385–98.

Lyde, L.W. (1915) *Some Frontiers of Tommorrow: An Aspiration for Europe* (London).

Lyde, L.W. (1916) 'River Frontiers in Europe', *Scottish Geographical Magazine* 32: 545–55.

Lyde, L.W. (1916–17) 'Europe v. Middle Europe', *Sociological Review* 9: 88–93.

Lyde, L.W. (1919) 'The International Rivers of Europe', *Geographical Journal* 54, 5: 303–13.

Lyde, L.W. (1931) *Peninsular Europe: Some Geographical Peregrinations, Ancient and Modern* (London).

Lynch, F.M.B. (1993) 'Restoring France: The Road to Integration', in A.S. Milward, F.M.B. Lynch, R. Raniero, F. Romero and V. Sørensen, *The Frontier of National Sovereignty: History and Theory 1945–1992* (London), 59–87.

Lyttleton, A. (1973) *The Seizure of Power: Fascism in Italy 1919–1929* (London).

McCauley, M. (1995) [1983] *The Origins of the Cold War 1941–1949* (London).

Macdougall, W.A. (1978) *France's Rhineland Diplomacy, 1914–24: The Last Bid for a Balance of Power in Europe* (Princeton).

McElligott, A. (1994) 'Reforging Mitteleuropa in the Crucible of War: The Economic Impact of Integration under German Hegemony', in P.M.R. Stirk (ed.), *Mitteleuropa: History and Prospects* (Edinburgh), 129–59.

Mackay, R. (1940) *Federal Europe* (London).

Mackenzie, J.M. (1984) *Propaganda and Empire: The Manipulation of British Public Opinion, 1880–1960* (Manchester)

Mackenzie, J.M. (1995) *Orientalism: History, Theory and the Arts* (Manchester).

Mackinder, H.J. (1887) 'On the Scope and Methods of Geography', *Proceedings of the Royal Geographical Society* 9: 141–60.

Mackinder, H.J. (1901) *Britain and the British Seas* (London).

Mackinder, H.J. (1904) 'The Geographical Pivot of History', *Geographical Journal* 23: 421–42.

Mackinder, H.J. (1905) 'Man-power as a Measure of National and Imperial Strength', *National and English Review* 45: 136–43.

Mackinder, H.J. (1911) 'The Teaching of Geography from an Imperial Point of View, and the Use which Could and Should be Made of Visual Instruction', *Geographical Teacher* 6: 79–86.

Mackinder, H.J. (1917) 'Some Geographical Aspects of International Reconstruction', *Scottish Geographical Magazine* 33: 1–11.

Mackinder, H.J. (1919) *Democratic Ideals and Reality: A Study in the Politics of Reconstruction* (London).

Mackinder, H.J. (1943) 'The Round World and the Winning of the Peace', *Foreign Affairs* 21: 595–605.

McLaren, A. (1983) *Sexuality and the Social Order: The Debate over the Fertility and Women and Workers in France 1770–1920* (New York).

McMillan, J.F. (1992) [1985], *Twentieth-Century France: Politics and Society 1898–1991* (London).

McNeil, W.H. (1964) *Europe's Steppe Frontier, 1500–1800* (Chicago).

McNeil, W.H. (1974) *The Shape of European History* (Oxford).

Malcolm, N. (1989) 'The "Common European Home" and Soviet European policy', *International Affairs* 65, 4: 659–76.

Mann, M. (1984) 'The Autonomous Power of the State: Its Origins, Mechanisms and Results', *Archives of European Sociology*, 25: 185–213.

Mann, M. (1986) *The Sources of Social Power*. Vol. 1 *A History of Power from the Beginning to AD 1760* (Cambridge).

Mann, M. (1993) *The Sources of Social Power*. Vol. 2 *The Rise of Classes and Nation-States, 1740–1914* (Cambridge).

Mann, T. (1935) *Achtung Europa* (Munich).

Manners, I.R. (1997) 'Constructing the Image of the City: The Representation of Constantinople in Christopher Buondelmonti's *Liber Insularum Archipelagi*', *Annals of the Association of American Geographers*, 87, 1: 72–102.

Margadant, T.W. (1992) *Urban Rivalries in the French Revolution* (Princeton).

Marquand, D. (1994) 'Re-inventing Federalism: Europe and the Left', *New Left Review* 203: 17–26.

Marrus, M.R. and Paxton, R.O. (1981) *Vichy France and the Jews* (Stanford).

Marshall, P.J. and Williams, G. (1982) *The Great Map of Mankind: British Perceptions of the World in the Age of the Enlightenment* (London).

Martin, G.J. (1980) *The Life and Thought of Isaiah Bowman* (Hamden, CT).

Marwick, A. (1964) 'Middle Option in the Thirties: Planning, Progress and Political "Agreement"', *The English Historical Review* 111: 285–98.

Masaryk, T.G. (1918) *The New Europe* (London).

Massie, R.K. (1991) *Dreadnought: Britain, Germany, and the Coming of the Great War* (New York).

Mastny, V. (1979) *Russia's Road to the Cold War: Diplomacy, Warfare and the Politics of Communism 1941–1945* (New York).

Matless, D. (1992) 'Regional Surveys and Local Knowledges: The Geographical Imagination in Britain, 1918–39', *Transactions of the Institute of British Geographers* New Series 17: 464–80.

Maurseth, P. (1964) 'Balance of Power Thinking from the Renaissance to the French Revolution', *Journal of Peace Research* 1, 2: 120–36.

Mayne, R., Pinder, J. with Roberts, J.C. de V. (1990) *Federal Union: The Pioneers* (London).

Mearsheimer, J.J. (1990) 'Back to the Future: Instability in Europe after the Cold War', *International Security* 15: 5–56.

Mee, C.L. (1984) *The Marshall Plan* (New York).

Meehan, E. (1993) *Citizenship and the European Community* (London).

Mellor, H. (1990) *Patrick Geddes: Social Evolutionist and City Planner* (London).

Mendras, H. (1997) *L'Europe des Européens: Sociologie de l'Europe occidentale* (Paris).

Mény, Y. (1974) *Centralisation and décentralisation dans le débat politique français (1945–1969)* (Paris).

Meny, Y. and Wright, V. (eds) (1985) *Centre–Periphery Relations in Western Europe* (London).

Merriman, J. (1996) *A History of Modern Europe*. Vol. 1. *From the Renaissance to the Age of Napoleon* (New York).

Mestrovic, S. (1993) *The Barbarian Temperament: Towards a Postmodern Critical Theory* (London).

Mestrovic, S. (1994) *The Balkanization of the West: the Confluence of Postmodernism and Postcommunism* (London).

Meyer, H.C. (1946) 'Mitteleuropa in German Political Geography', *Annals of the Association of American Geographers* 36: 178–94.

Meyer, H.C. (1955) *Mitteleuropa in German Thought and Practice 1815–1945* (The Hague).

Meyer, J.W. (1989) 'Conceptions of Christendom: Notes on the Distinctiveness of the West', in M.L. Kohn (ed.), *Cross National Research in Sociology* (London) 74–85.

Miall, H. (1993) *Shaping the New Europe* (London).

Miall, H. (ed.) (1994) *Minority Rights in Europe* (London).

Michalka, W. (1985) 'From the Anti-Comintern Pact to the Euro-Asiatic Bloc: Ribbentrop's Alternative Concept of Hitler's Foreign Policy Programme', in H.W. Koch (ed.), *Aspects of the Third Reich* (London), 267–84.

Midlarsky, M. (1975) *On War: Political Violence in the International System* (New York).

Midlarsky, M. (1981) 'Equilibria in the Nineteenth-Century Balance of Power System', *American Journal of Political Science* 25: 270–96.

Miles, R. (1994) 'Explaining Racism in Contemporary Europe', in A. Rattansi and S. Westwood (eds), *Racism, Modernity and Identity on the Western Front* (Cambridge), 189–221.

Milward, A.S. (1984) *The Reconstruction of Western Europe, 1945–1951* (London).

Milward, A.S. (1992) *The European Rescue of the Nation-State* (London).

Milward, A.S. (1993) 'Conclusions: The Value of History', in A.S. Milward, F.M.B. Lynch, F. Romero, R. Raniero and V. Sørensen (1993) *The*

Frontier of National Sovereignty: History and Theory 1945–1992 (London), 182–201.

Milward, A.S. and Sørensen, V. (1993) 'Interdependence of Integration? A National Choice', in A.S. Milward, F.M.B. Lynch, F. Romero, R. Raniero and V. Sørensen (1993) *The Frontier of National Sovereignty: History and Theory 1945–1992* (London), 1–32.

Mitchell, P.C. (1915) *Evolution and the War* (London).

Mlinar, Z. (ed.) (1992) *Globalization and Territorial Identities* (Aldershot).

Mommsen, H. (1963) *Die Socialdemokratie und die Nationalitätenfrage im habsburgischen Vielvölkerstaat* (Vienna).

Moon, D. (1997) 'Peasant Migration and the Settlement of Russia's Frontiers, 1550–1897', *Historical Journal* 40, 4: 859–93.

Moravcsik, A. (1991) 'Negotiating the Single European Act: National Interests and Conventional Statecraft in the European Community', *International Organization* 45: 19–56.

Morel, E.D. (1916) *Truth and the War* (London).

Morgan, P.J. (1990) 'The Italian Fascist New Order in Europe', in M.L. Smith and P.M.R. Stirk (eds), *Making the New Europe: European Unity and the Second World War* (London), 27–45.

Morgan, P.J. (1996) '"A Vague and Puzzling Idealism...": Plans for European Unity in the Era of the Modern State', in M. Wintle (ed.), *Culture and Identity in Europe: Perceptions of Divergence and Unity in Past and Present* (Aldershot), 33–51.

Morin, E. (1991) *Concepts of Europe* (New York).

Mosse, G.L. (1978) *Toward the Final Solution: A History of European Racism* (New York).

Mosse, G.L. (1980) *Masses and Man: Nationalist and Fascist Perceptions of Reality* (New York).

Mosse, G.L. (1981) [1964], *The Crisis of German Ideology: The Intellectual Origins of the Third Reich* (New York).

Mosse, G.L. (1990) *Fallen Soldiers: Reshaping the Memory of the World Wars* (Oxford).

Moul, W. (1985) 'Balances of Power and European Great Power War, 1815–1939: A Suggestion and Some Evidence', *Canadian Journal of Political Science* 43: 481–528.

Mukerji, C. (1997) *Territorial Ambitions and the Gardens of Versailles* (Cambridge).

Müller-Hill, B. (1988) *Murderous Science: Elimination by Scientific Selection of Jews, Gypsies and Others in Germany, 1933–1945* (Oxford).

Munro, R. (1919) *From Darwinism to Kaiserism, being a Review of the Origin, Effects and Collapse of Germany's Attempt at World Dominion by Methods of Barbarism* (Glasgow).

Murphy, A.B. (1990) 'Historical Justifications of Territorial Claims', *Annals of the Association of American Geographers* 80: 531–48.

Najam, E.W. (1956) 'Richelieu's Blueprint for Unity and Peace', *Studies in Philology*, 53: 25–34.

Nalkowski, W. (1917) [1912] *Poland as a Geographical Entity* (London).

Nathan, J.A. (1980) 'The Heyday of the Balance of Power: Frederick the Great and the Decline of the Old Regime', *Naval War College Review* 33: 53–67.

Naumann, F. (1916) [1915] *Central Europe* (London).

Navari, C. (1991) 'The Origins of the Briand Plan', in A. Bosco (ed.), *The Federal Idea* Vol. 1 (London), 210–35.

Nederveen Pieterse, J. (1991) 'Fictions of Europe', *Race and Class*, 32, 3: 3–10

Nederveen Pieterse, J. (1994) 'Unpacking the West: How European is Europe?', in A. Rattensi and S. Westwood (eds), *Racism, Modernity and Identity on the Western Front* (Cambridge), 129–49.

Nelson, B.F. and Stubb, A.C.G. (eds) (1994) *The European Union: Readings on the Theory and Practice of European Integration* (Boulder, CA).

Nelson, B.F., Roberts, D. and Veit, W. (eds) (1992) *The Idea of Europe: Problems of National and Transnational Identity* (Oxford).

Nelson, E.W. (1943) 'The Origins of Modern Balance of Power Politics', *Medievalia et Humanistica* 1: 124–42.

Neumann, I.B. (1993) 'Russia as Central Europe's Constituting Other', *East European Politics and Societies* 7: 348–69.

Neumann, I.B. (1996) *Russia and the Idea of Europe: A Study in Identity and International Relations* (London).

Neumann, I.B. and Welsh, J.M. (1991) 'The Other in European Self-Definition: An Addendum to the Literature on International Society', *Review of International Studies* 7: 327–48.

Newbigin, M.I. (1915) *Geographical Aspects of the Balkan Problems in Their Relation to the Great European War* (London).

Newbigin, M.I. (1917) 'Race and Nationality', *Geographical Journal* 50, 5: 313–29.

Newbigin, M.I. (1918) 'The Origin and Maintenance of Diversity in Man', *Geographical Review* 6: 411–20.

Newbigin, M.I. (1919) 'The Polish Problem', *Scottish Geographical Magazine* 35: 81–93.

Newbigin, M.I. (1920) *Aftermath: The Geographical Study of the Peace Terms* (Edinburgh).

Newman, M. (1983) *Socialism and European Unity* (London).

Nicholas, L.H. (1995) *The Rape of Europa: The Fate of Europe's Treasures in the Third Reich and the Second World War* (New York).

Nisbet, R. (1980) *History of the Idea of Progress* (London).

Noakes, J., Pridham, G. (1988) *Nazism, 1919–1945. Vol III – Foreign Policy, War and Racial Extermination* (Exeter).

Noel-Baker, P. (ed.) (1934) *Challenge to Death* (London).

Nolte, E. (1965) [1963] *Three Faces of Fascism: Action Française, Italian Fascism, and National Socialism* (London).

Nolte, E. (1968) *Die Krise des liberalen Systems und die faschistischen Bewegungen* (Munich).

Nolte, E. (1987) *Der europaische Burgerkrieg 1917–1945: Nationalsozialismus und Bolschewismus* (Berlin).

Nordman, D. (1986) 'Des limites d'État aux frontières nationales', in P. Nora (ed.), *Les lieux de mémoire*. Vol. 2, Part 2. *La nation* (Paris), 35–61.

Nordman, D. and Revel, J. (1989) 'La formation de l'espace français', in A. Burgière and J. Revel (eds), *Histoire de la France*. Vol. 1 (Paris).

Nordman, D. and Vic-Ozouf Marignier, M. (1989a), *Atlas de la Révolution française*. Vol. 4 – *Le territoire: réalités et représentations* (Paris).

Nordman, D. and Vic-Ozouf Marignier, M. (1989b), *Atlas de la Révolution française*. Vol. 5 – *Le territoire et les limites administratives* (Paris).

Norton, D. (1968) 'Karl Haushofer and the Germany Academy, 1925–1945', *Central European History* 1: 80–99.

Nuttal, E.M. (1927) *A Project for Perpetual Peace: Rousseau's Essay* (London).

Nye, R.A. (1984) *Madness and Politics in Modern France: The Medical Concept of National Decline* (Princeton).

O'Brien, R. (1992) *Global Financial Integration: The End of Geography* (London).

Offen, K. (1984) 'Depopulation, Nationalism and Feminism in Fin-de-Siècle France', *American Historical Review* 89: 648–75.

Ogden, P. and Huss, M.-M. (1982) 'Demography and Pronatalism in France in the Nineteenth and Twentieth Centuries', *Journal of Historical Geography* 8: 283–96.

Ogg, D. (ed.) (1921) *Sully's Grand Design of Henry IV: From the Memoirs of Maximilien de Béthune, Duc de Sully (1559–1641)* (London).

Ogilvie, A.G. (1922) *Some Aspects of Boundary Settlement at the Peace Conference* (London).

Ohmae, K. (1990) *The Borderless World: Power and Strategy in the Inter-Linked Economy* (London).

Ohmae, K. (1995) *The End of the Nation-State: The Rise of Regional Economies* (London).

Okey, R. (1992) 'Central Europe/Eastern Europe: Behind the Definitions', *Past and Present* 137: 102–33.

Olin, M. (1997) 'Lanzmann's *Shoah* and the Topography of the Holocaust Film', *Representations* 57: 1–23.

O'Loughlin, J. and van der Wusten, H. (1990) 'The Political Geography of Panregions', *Geographical Review* 80, 1: 1–20.

O'Loughlin, J. and van der Wusten, H. (eds) (1993) *The New Political Geography of Eastern Europe* (London).

O'Loughlin, J., Flint, C. and Anselin, L. (1994) 'The Geography of the Nazi

Vote: Context, Confession, and Class in the Reichstag Election of 1930',
Annals of the Association of American Geographers 84: 351–80.

O'Loughlin, J., Flint, C. and Shin, M. (1995) 'Regions and Milieux in
Weimar Germany: The Nazi Vote of 1930 in Geographic Perspective',
Erdkunde 49: 305–14.

Ormsby, H. (1935) 'The Definition of Mitteleuropa and Its Relation to the
Concept of Deutschland in the Writings of Modern German
Geographers', *Scottish Geographical Magazine* 51: 337–47.

Ortega Y Gasset, J. (1930) *The Revolt of the Masses* (London).

O'Sullivan, P. (1986) *Geopolitics* (New York).

Ó Tuathail, G. (1996) *Critical Geopolitics* (London).

Ó Tuathail, G. and Agnew, J. (1992) 'Geopolitics and Discourse: Practical
Geopolitical Reasoning in American Foreign Policy', *Political Geography*
11: 190–204.

Ó Tuathail, G., Dalby, S. and Routledge, P. (eds) (1998) *The Geopolitics
Reader* (London).

Outram, D. (1995) *The Enlightenment* (Cambridge).

Overy, R.J. (1982) *The Nazi Economic Recovery 1932–1938* (London).

Ozouf, M. (1988) *Festivals and the French Revolution* (Cambridge, MA).

Paasi, A. (1995) *Territories, Boundaries and Consciousness: The Changing
Geographies of the Finnish–Russian Border* (Chichester).

Painter, J. (1995) *Politics, Geography and 'Political Geography'* (London).

Pajic, Z. (1993) 'The Structures of Apartheid: The New Europe of Ethnic
Division', *War Report* 21: 3–4.

Pallot, J. and Shaw, D.J.B. (1990) *Landscape and Settlement in Romanov
Russia 1613–1917* (Oxford).

Palmer, A. (1970) *The Lands Between: A History of East-Central Europe
since the Congress of Vienna* (London).

Palmer, J. (1987) *Europe Without America* (London).

Parker, G. (1969) *The Logic of Unity: A Geography of the European
Economic Community* (London).

Parker, G. (1985) *Western Geopolitical Thought in the Twentieth Century*
(London).

Parker, G. (1987) 'French Geopolitical Thought in the Inter-War Years and
the Emergence of the European Idea', *Political Geography Quarterly* 6:
145–50.

Parker, G. (1988) *The Geopolitics of Domination* (London).

Parker, G. (ed.) (1995) *Cambridge Illustrated History of Warfare* (Cambridge).

Parker, G. and Smith, L. (eds) (1978) *The General Crisis of the Seventeenth
Century* (London).

Parker, W.H. (1960) 'Europe: How Far?', *Geographical Journal* 126:
278–97.

Parker, W.H. (1968) *An Historical Geography of Russia* (London).

Parker, W.H. (1982) *Mackinder: Geography as an Aid to Statecraft*
(Oxford).

Partsch, J. (1904) *Mitteleuropa: Due Länder und Völker von den Westalpen und dem Balkan bis an den Kanal und das kurische Haff* (Gotha).

Paul, D.B. (1984) 'Eugenics and the Left', *Journal of the History of Ideas* 45: 567–90.

Paxton, R.O. (1972) *The Vichy Regime: Old Guard and New Order* (London).

Payne, S.G. (1995) *A History of Fascism 1914–1945* (London).

Peake, H.J. (1919) 'Devolution: A Regional Movement. A: Provinces of England. B: European Aspects', *Sociological Review* 11: 97–113.

Pearson, R. (1995) 'The Making of '89: Nationalism and the Dissolution of Communist Eastern Europe', *Nations and Nationalism* 1, 1: 69–75.

Pecqueur, C. (1842) *De la paix, de son principe et de sa réalisation* (Paris).

Peet, R. (1985) 'The Social Origins of Environmental Determinism', *Annals of the Association of American Geographers* 75, 3: 309–33.

Pegg, C.H. (1983) *Evolution of the European Idea, 1914–1932* (Chapel Hill).

Pelletier, M. (1990) *La carte de Cassini: l'extraordinaire aventure de la carte de France* (Paris).

Penck, A. (1915) 'Politisch-geographische Lehren des Krieges', *Meereskunde* 9–10: 12–21.

Penck, A. (1916) 'Der Krieg und das Studium der Geographie', *Zeitschrift der Gesellschaft für Erdkunde zu Berlin* 159–76 and 222–48.

Penck, A. (1925) 'Deutscher Volks- und Kulturboden', in K.C. von Loesch and A.H. Ziegfeld (eds), *Volk unter Völkern* (Breslau), 62–85.

Penck, A. and Fischer, H. (1925) *Der deutsche Volks- und Kulturboden in Europa* (Berlin).

Penn, W. (1936) [1693] *An Essay Towards the Present and Future Peace of Europe by the Establishment of an European Diet, Parliament, or Estates* (London).

Perkins, M.L. (1959) *The Moral and Political Philosophy of the Abbé de Saint-Pierre* (Geneva and Paris).

Perrot, J.-C. and Woolf, S. (1984) *State and Statistics in France 1789–1815* (London).

Pesonen, P. (1991) 'The Image of Europe in Russian Literature and Culture', *History of European Ideas* 13, 4: 399–409.

Pick, D. (1989) *Faces of Degeneration: A European Disorder, c. 1848–c. 1918* (Cambridge).

Pick, D. (1993) *War Machine: The Rationalisation of Slaughter in the Modern Age* (New Haven).

Pinder, J. (1986) 'Federal Union 1939–41', in W. Lipgens (ed.), *Documents on the History of European Integration.* vol. 2, 26–34.

Pinder, J. (1989) 'Federalism in Britain and Italy: Radicals and the English Liberal Tradition', in P.M.R. Stirk (ed.), *European Unity in Context: The Interwar Period* (London), 201–23.

Piveteau, J.-L. (1995) *Temps du territoire* (Paris).

Pinder, J. (1991) *The European Community and Eastern Europe* (London).

Pocock, J.G.A. (1975) *The Machiavelli Moment: Florentine Political Thought and the Atlantic Republican Tradition* (Princeton).

Pocock, J.G.A. (1991) 'Deconstructing Europe', *London Review of Books* 13, 24: 6–10.

Pogge, T.W. (1992) 'Cosmopolitanism and Sovereignty', *Ethics* 103: 48–75.

Poliakov, L. (1974) *The Aryan Myth: A History of Racist and Nationalist Ideas in Europe* (London).

Pollard, M. (1998) *Reign of Virtue: Mobilising Gender in Vichy France* (Chicago).

Popovski, V. (1995) 'Yugoslavia: Politics, Federation, Nation', in G. Smith (ed.), *Federalism: The Muti-Ethnic Challenge* (London), 180–207.

Pounds, N.J.G. (1951) 'The Origins of the Idea of Natural Frontiers in France', *Annals of the Association of American Geographers* 41: 146–57.

Pounds, N.J.G. (1954) 'France and "les limites naturelles" from the Seventeenth to the Twentieth centuries', *Annals of the Association of American Geographers* 44: 51–62.

Pounds, N.J.G. (1979) *An Historical Geography of Europe 1500–1840* (Cambridge).

Pounds, N.J.G. (1985) *An Historical Geography of Europe 1800–1914* (Cambridge).

Pounds, N.J.G. and Ball, S.S. (1964) 'Core Areas and the Development of the European State System', *Annals of the Association of American Geographers*, 54: 24–40.

Prescott, J.R.V. (1978) [1965] *The Geography of Frontiers and Boundaries* (London).

Proudhon, P.-J. (1979) [1863], *The Principle of Federation* (Toronto).

Prucha, V. (1990) 'The Integration of Czechoslovakia in the Economic System of Nazi Germany', in M.L. Smith and P.M.R. Stirk (eds), *Making the New Europe: European Unity and the Second World War* (London), 87–97.

Raffestin, C., Lopreno, D. and Pasteur, Y. (1995) *Géopolitique et histoire* (Lausanne).

Ranieri, R. (1991) 'Attempting an Unlikely Union: The British Steel Industry and the European Coal and Steel Community, 1950–54', in P.M.R. Stirk and D. Willis (eds), *Shaping Postwar Europe: European Unity and Disunity 1945–1957* (London), 112–23.

Ratzel, F. (1897) *Politische Geographie oder die Geographie der Staaten, des Verkehres und des Krieges* (Munich/Leipzig).

Ratzel, F. (1900) *Das Meer als Quelle des Völkergrösse. Eine politische-geographische Studie* (Munich).

Ratzel, F. (1901) 'Der Lebensraum. Eine biogeographische Studie', in K. Bücher and K. Fricker (eds), *Festgaben für Albert Schäffle zur siebenzigsten Wiederkehr sienes Geburtstags am 24. Februar. 1901* (Tübingen) 101–90.

Reclus, O. (1914) *L'Allemagne en morceaux: paix draconnienne* (Paris).

Renan, E. (1990) [1882], 'What is a Nation?', in H.K. Bhabha (ed.), *Nation and Narration* (London), 8–21.

Renouvin, P. (1949) *L'Idée de fédération européenne dans la pensée politique du XIXe. siècle* (Oxford).

Reuter, T. (1992) 'Medieval Ideas on Europe and their Modern Historians', *History Workshop Journal* 33: 176–80.

Rich, P. (1984) '"The Baptism of a New Era": the 1911 Universal Races Congress and the Liberal Ideology of Race', *Ethnic and Racial Studies* 7, 4: 534–50.

Richards, T. (1993) *The Imperial Archive: Knowledge and the Fantasy of Empire* (London).

Ridley, F.A. and Edwards, B. (1944) *The United Socialist States of Europe* (London).

Ripley, W.Z. (1899) *The Races of Europe: A Sociological Study* (New York).

Roberts, J.M. (1985) *The Triumph of the West* (London).

Robic, M.-C. (1994) 'National Identity in Vidal's *Tableau de la géographie de la France*: From Political Geography to Human Geography', in D. Hooson (ed.), *Geography and National Identity* (Oxford), 58–70.

Robic, M.-C. (1996) 'Les vœux des premiers Congrès: dresser la Carte du Monde', in M.C. Robic, A.-M. Briend and M. Rössler (eds), *Géographes face au monde* (Paris), 149–78.

Rokkan, S. and Urwin, D.W. (eds) (1982) *The Politics of Territorial Identity: Studies in European Regionalism* (London).

Rollins, W.H. (1995) 'Whose Landscape? Technology, Fascism, and Environmentalism on the Nationalist Socialist Autobahn', *Annals of the Association of American Geographers* 85, 3: 494–520.

Romero, F. (1993) 'Interdependence and Integration in American Eyes: From the Marshall Plan to Currency Convertibility', in A.S. Milward, F.M.B. Lynch, F. Romero, R. Raniero and V. Sørensen, *The Frontier of National Sovereignty: History and Theory 1945–1992* (London), 155–81.

Romm, J.S. (1992) *The Edges of the Earth in Ancient Thought* (Princeton).

Roobol, W.H. (1990) 'The Prospects for the German Domination of Europe in the Era of the World Wars', in M.L. Smith and P.M.R. Stirk (eds), *Making the New Europe: European Unity and the Second World War* (London), 18–26.

Roosevelt, G.G. (1990) *Reading Rousseau in the Nuclear Age* (Philadelphia).

Rosenberg, A. (1927) *Der Zukunftsweg einer deutschen Aussenpolitik* (Munich).

Ross, G. (1983) *The Great Powers and the Decline of the European States System 1914–45* (London).

Rössler, M. (1989) 'Applied Geography and Area Research in Nazi Society:

Central Place Theory and Planning, 1933 to 1945', *Environment and Planning D: Society and Space* 74: 419–31.

Rössler, M. (1994) 'Berlin or Bonn? National Identity and the Question of the German Capital', in D. Hooson (ed.), *Geography and National Identity* (Oxford), 92–103.

Rostow, W.W. (1960) *The Stages of Economic Growth: A Non-Communist Manifesto* (Cambridge).

Rousseau, J.-J. (1761) *Extrait du projet de paix perpétuelle de Monsieur l'Abbé de Saint-Pierre* (Amsterdam).

Rubin, M. (1992) 'The Culture of Europe in the Later Middle Ages', *History Workshop Journal* 33: 162–75.

Ruggie, J.G. (1993) 'Territoriality and Beyond: Problematizing Modernity in International Relations', *International Organization* 47: 139–74.

Rupnik, J. (1990) 'Central Europe or Mitteleuropa?', *Daedalus* 119, 1: 249–78.

Ryan, H.B. (1982) *The Vision of Anglo–America* (Cambridge).

Ryan, J. (1994) 'Visualizing Imperial Geography: Halford Mackinder and the Colonial Office Visual Instruction Committee, 1902–1911', *Ecumene* 1: 157–76.

Sack, R.D. (1980) *Conceptions of Space in Social Thought: A Geographic Perspective* (London)

Sahlins, P. (1990) 'Natural Frontiers Revisited: France's Boundaries since the Seventeenth Century', *American Historical Review* 95: 1423–51.

Said, E.W. (1978) *Orientalism* (London).

Said, E.W. (1993) *Culture and Imperialism* (London).

Salewski, M. (1985) 'Ideas of the National Socialist Government and Party', in W. Lipgens (ed.), *Documents on the History of European Integration* Vol. 1, 37–54.

Sandholtz, W. and Zysman, J. (1989) '1992: Recasting the European Bargain', *World Politics* 41: 95–128.

Sandner, G. (1989) 'The *Germania triumphans* Syndrome and Passarge's *Erdkundliche Weltanschauung*: The Roots and Effects of German Political Geography beyond *Geopolitik*', *Political Geography Quarterly* 8, 4: 341–51.

Schama, S. (1996) *Landscape and Memory* (London).

Scheler, M. (1915) *Der Genius des Krieges und der deutsche Krieg* (Berlin).

Schlaim, A. (1974) 'Prelude to Downfall: The British Offer of Union to France, June 1940', *Journal of Contemporary History* 9: 27–63.

Schmidt, H.D. (1966) 'The Establishment of "Europe" as a Political Expression', *The Historical Journal* 9, 2: 172–8.

Schmitt, C. (1941) *Völkerrechtliche Grossraumordnung* (Berlin).

Schöpflin, G. and Wood, N. (eds) (1989) *In Search of Central Europe* (Oxford).

Schorske, C.E. (1979) [1961] *Fin-de-Siècle Vienna* (New York).

Schöttler, P. (1995) 'The Rhine as an Object of Historical Controversy in the Inter-War Years: Towards a History of Frontier Mentalities', *History Workshop Journal* 39: 1–21.

Schöttler, P. (1997) 'Lucien Febvre ou la démystification de l'histoire rhénane', preface to L. Febvre, *Le Rhin: histoire, mythes et réalités* (Paris), 11–56 [a reprint of Febvre's section in Demangeon and Febvre (1935)].

Schroeder, P. (1989) 'The Nineteenth-Century System: Balance of Power or Political Equilibrium?', *Review of International Studies* 15: 135–53.

Schultz, H.D. (1989) 'Fantasies of *Mitte*: *Mittelage* and *Mitteleuropa* in German Geographical Discussion in the 19th and 20th Centuries', *Political Geography Quarterly* 8: 315–40.

Scott, A., Peterson, J. and Millar, D. (1994) 'Subsidiarity: A "Europe of the Regions" v. the British Constitution', *Journal of Common Market Studies* 32: 47–67.

Sereny, G. (1995) *Albert Speer: His Battle With the Truth* (London).

Servan-Schreiber, J.-J. (1967) *Le défi américain* (Paris).

Seton-Watson, H. (1985) 'What is Europe? Where is Europe? From Mystique to Politique', *Encounter*, 60, 2: 9–17.

Seton-Watson, R.W. (1917) *The Rise of Nationality in the Balkans* (London).

Seymour, C. (1951) *Geography, Justice and Politics at the Paris Peace Conference* (New York).

Shand, J.D. (1984) 'The *Reichsautobahn*: Symbol of the Third Reich', *Journal of Contemporary History* 19, 2: 191–6.

Sharp, A. (1991) *The Versailles Settlement: Peacemaking in Paris, 1919* (London).

Shaw, D.J.B. (1996) 'Geographical Practice and its Significance in Peter the Great's Russia', *Journal of Historical Geography* 22, 2: 160–76.

Sheehan, M. (1988) 'British Thinking on the Balance of Power 1660–1714', *History* 73: 24–37.

Sheehan, M. (1996) *The Balance of Power: History and Theory* (London).

Shennan, J.H. (1974) *The Origins of the Modern European State 1450–1725* (London).

Sidaway, J. (1994) 'Political Geography in the Time of Cyberspace: New Agendas?', *Geoforum* 25: 487–503.

Silber, L. and Little, A. (1996) *The Death of Yugoslavia* (London).

Silverman, M. (1992) *Deconstructing the Nation: Immigration, Racism and Citizenship in Modern France* (London).

Simon, E.D. (1939) *The Smaller Democracies of Europe* (London).

Sinnhuber, K.A. (1954) 'Central Europe – Mitteleuropa – Europe Centrale', *Transactions and Papers of the Institute of British Geographers* 20: 15–39.

Skinner, Q. (1981) *Machiavelli* (Oxford).

Slezkine, Y. (1994) *Artic Mirrors: Russia and the Small Peoples of the North* (Ithaca).

Smith, A.D. (1981) 'War and Ethnicity: The Role of Warfare in the Formation, Self-Images and Cohesion of Ethnic Communities', *Ethnic and Racial Studies* 4, 4: 375–97.

Smith, A.D. (1986) *The Ethnic Origins of Nations* (Oxford).

Smith, A.D. (1991) *National Identity* (Harmondsworth).

Smith, A.D. (1992) 'National Identity and the Idea of European Unity', *International Affairs* 68: 55–76.

Smith, C.T. (1978) [1967] *An Historical Geography of Western Europe before 1800* (London).

Smith, G. (ed.) (1995) *The Nationalities Question in the Post-Soviet States* (London).

Smith, M. (1984) *Western Europe and the United States: The Uncertain Alliance* (London).

Smith, M.L. (1989) 'Ideas of a New Order in France, Britain and the Low Countries in the 1930s', in P.M.R. Stirk (ed.), *European Unity in Context: The Interwar Period* (London), 149–69.

Smith, M.L. (1990) 'The Anti-Bolshevik Crusade and Europe', in M.L. Smith and P.M.R. Stirk (eds), *Making the New Europe: European Unity and the Second World War* (London), 46–65.

Smith, N. (1969) 'The Idea of the French Hexagon', *French Historical Studies* 3: 139–55.

Smith, N. (1984) 'Isaiah Bowman: Political Geography and Geopolitics', *Political Geography Quarterly* 3: 69–76.

Smith, N. (1986) 'Bowman's New World and the Council on Foreign Relations', *Geographical Review* 76: 438–60.

Smith, N. (1994) 'Shaking Loose the Colonies: Isaiah Bowman and the "Decolonisation" of the British Empire', in A. Godlewska and N. Smith (eds), *Geography and Empire* (Oxford), 270–99.

Smith, W.D. (1980) 'Friedrich Ratzel and the Origins of *Lebensraum*', *German Studies Review* 3: 51–68.

Smith, W.D. (1986) *The Ideological Origins of Nazi Imperialism* (Oxford).

Soja, E. (1989) *Postmodern Geographies: The Reassertion of Space in Critical Social Theory* (London).

Soueleyman, E. (1941) *The Vision of World Peace in Seventeenth-Century France* (New York).

Southgate, B.C. (1993) '"Scattered over Europe": Transcending National Frontiers in the Seventeenth Century', *History of European Ideas*, 16: 131–7.

Spackman, B. (1996) Fascist Virilities: Rhetoric, Ideology and Social Fantasy in Italy (Minneapolis).

Spengler, J.-J. (1979) *France Faces Depopulation* (Westport, CT).

Spengler, O. (1918–22) *The Decline of the West* (2 vols, London).

Spicer, M. (1992) *A Treaty Too Far: A New Policy for Europe* (London).

Springborg, P. (1992) *Western Republicanism and the Oriental Prince* (Cambridge).

Spykman, N. (1944) *The Geography of the Peace* (New York).

Stall, R.J. (1984) 'Bloc Concentration and the Balance of Power: The European Major Powers 1824–1914', *Journal of Conflict Resolution* 28: 25–50.

Staszak, J.-F. (1995) *La géographie avant la géographie: le climat chez Aristote et Hippocrate* (Paris).

Steinberg, J. (1990) *All or Nothing: The Axis and the Holocaust 1941–43* (London).

Stepan, N.L. (1982) *The Idea of Race in Science: Great Britain, 1800–1960* (London).

Stepan, N.L. (1987) '"Nature's Pruning Hook": War, Race and Evolution, 1914–18', in J.M.W. Bean (ed.), *The Political Culture of Modern Britain: Studies in the Memory of Stephen Koss* (London), 129–48.

Stephan, J.J. (1978) *The Russian Fascists: Tragedy and Farce in Exile 1925–1945* (New York).

Stevenson, D. (1982) *French War Aims against Germany 1914–1919* (Oxford).

Stirk, P.M.R. (1989) 'Authoritarian and Nationalist Socialist Conceptions of Nation, State and Europe', in P.M.R. Stirk (ed.), *European Unity in Context: The Interwar Period* (London), 125–48.

Stirk, P.M.R. (1990) 'Anti-Americanism in National Socialist Propaganda during the Second World War', in M.L. Smith and P.M.R. Stirk (eds), *Making the New Europe: European Unity and the Second World War* (London), 66–86.

Stirk, P.M.R. (1991) 'Americanism and Anti-Americanism in British and German Responses to the Marshall Plan', in P.M.R. Stirk and D. Willis (eds), *Shaping Postwar Europe: European Unity and Disunity 1945–1957* (London), 27–42.

Stirk, P.M.R. (1994a), 'The Idea of Mitteleuropa', in P.M.R. Stirk (ed.), *Mitteleuropa: History and Prospects* (Edinburgh), 1–35.

Stirk, P.M.R. (1994b), 'Ideas of Economic Integration in Interwar Mitteleuropa', in P.M.R. Stirk (ed.), *Mitteleuropa: History and Prospects* (Edinburgh), 86–111.

Stirk, P.M.R. (1996) *A History of European Integration since 1914* (London).

Stocking, G.W. (1987) *Victorian Anthropology* (London).

Stoianovich, T. (1976) *French Historical Method: The Annales Paradigm* (Ithaca, NY).

Stokes, G. (1980) *Hitler and the Quest for World Domination* (Leamington Spa).

Strachey, J. (1940) *Federalism or Socialism* (London).

Strausz-Haupé, R. (1942) *Geopolitics: The Struggle for Space and Power* (New York).

Streit, C.K. (1939) *Union Now: A Proposal for a Federal Union of the Democracies of the North Atlantic* (London).

Streit, C.K. (1941) *Union Now With Britain* (London).

Strohmayer, U. (1996) 'Pictorial Symbolism in the Age of Innocence: Material Geographies at the Paris World's Fair of 1937', *Ecumene* 3, 3: 282–304.

Stromberg, R.N. (1982) *Redemption by War: The Intellectuals and 1914* (Lawrence, KS).

Sullivan, R. (1973) 'Machiavelli's Balance of Power Theory', *Social Science Quarterly* 54: 258–70.

Sumner, B.H. (1973) *Peter the Great and the Emergence of Russia* (New York).

Sutherland, D.M.G. (1985) *France 1789–1815: Revolution and Counter-Revolution* (London).

Swartz, M. (1971) *The Union of Democratic Control in British Politics during the First World War* (Oxford).

Szücs, J. (1988) [1983], 'Three historical regions of Europe', in J. Keane (ed.), *Civil Society and the State: New European Perspectives* (London), 291–332.

Talmor, E. (1980) 'Reflections on the Rise and Development of the Idea of Europe', *History of European Ideas*, 1: 63–6.

Taylor, A.J.P. (1963) *The First World War: An Illustrated History* (London).

Taylor, A.J.P. (1969) *War by Timetable: How the First World War Began* (London).

Taylor, G. (1946) *Our Evolving Civilization: An Introduction to Geopacifics – Geographical Aspects of the Path Towards World Peace* (Oxford).

Taylor, J.E. (1993) *Christians and Holy Places* (Oxford).

Taylor, K. (ed.) (1975) *Henri Saint-Simon (1760–1825): Selected Writings on Science, Industry and Social Organization* (London).

Taylor, P. (1983) *The Limits of European Integration* (London).

Taylor, P.J. (1990) *Britain and the Cold War: 1945 as Geopolitical Transition* (London).

Taylor, P.J. (1991) 'A Theory and Practice of Regions: The Case of Europe', *Environment and Planning D: Society and Space* 92: 183–95.

Taylor, P.J. (1993) [1985] *Political Geography: World Economy, Nation-State and Locality* 3rd edn (Harlow).

Taylor, P.J. (1994) 'The State as Container: Territoriality in the Modern World System', *Progress in Human Geography*, 18: 151–62.

Taylor, P.J. (1995) 'Beyond Containers: Internationality, Interstateness, Interterritoriality', *Progress in Human Geography*, 19: 1–15.

Taylor, P.J. (1996) *The Way the Modern World Works: World Hegemony to World Impasse* (Chichester).

Tazbir, J. (1977) 'Poland and the Concept of Europe in the Sixteenth–Eighteenth Centuries', *European Studies Review* 7: 29–45.

Teggart, F.J. (1919) 'Geography as an Aid to Statecraft: An Appreciation of

Mackinder's "Democratic Ideals and Reality"', *Geographical Review* 8: 227–42.

Teich, M. and Porter, R. (eds) (1990) *Fin-de-Siècle and its Legacy* (Cambridge).

Temperley, H.V. (1961) *A History of the Peace Conference of Paris* (4 Vols, London).

Thaden, E.C. (1984) *Russia's Western Borderlands, 1710–1870* (Princeton).

Thernborn, G. (1995) *European Modernity and Beyond: The Trajectory of European Societies 1945–2000* (London).

Thomas, H. (1977) *The Spanish Civil War* (Harmondsworth).

Tilly, C. (1975) *The Formation of Nation States in Western Europe* (Princeton).

Tilly, C. (1989) 'The Geography of European Statemaking and Capitalism since 1500', in E.D. Genovese and L. Hochberg (eds), *Geographic Perspectives in History* (Oxford), 158–81.

Trachtenberg, M. (1980) *Reparation in World Politics: France and European Economic Diplomacy 1916–23* (New York).

Trevor-Roper, H.R. (1973) *Hitler's Table Talk, 1941–1944* (London).

Tribe, K. (1978) *Land, Labour and Economic Discourse* (London).

Trotsky, L.D. (1971) [1926] *Europe and America: Two Speeches on Imperialism* (New York).

Tuan, Y.-F. (1996) *Cosmos and Hearth: A Cosmopolite's Viewpoint* (Minneapolis).

Turner, H.A. (1985) *German Big Business and the Rise of Hitler* (Oxford).

Turnock, D. (1989) *Eastern Europe: An Historical Geography* (London).

Turnock, D. (1997) *The Eastern European Economy in Context: Communism and Transition* (London).

Unstead, J.F. (1916) 'A Synthetic Method of Determining Geographical Regions', *Geographical Journal* 48, 3: 230–49.

Urwin, D.W. (1989) *Western Europe since 1945* (London).

Urwin, D.W. (1995) [1991] *The Community of Europe: A History of European Integration* (London).

Vagts, A. (1948) 'The Balance of Power: Growth of an Idea', *World Politics* 1: 82–101.

Vagts, A and Vagts, D. (1979) 'The Balance of Power in International Law: A History of an Idea', *American Journal of International Law* 73: 555–79.

Van der Vat, D. (1997) *The Good Nazi: The Life and Lies of Albert Speer* (London).

Van Ham, P. (1993) *The EC, Eastern Europe and European Unity* (London).

Vaughan, R. (1979) *Twentieth-Century Europe: Paths to Unity* (London).

Verosta, S. (1977) 'The German Concept of Mitteleuropa 1916–1918 and its Contemporary Critics', in R.A. Kann, B.K. Kiraly, and S. Fichtner (eds), *The Habsburg Empire in World War I* (Boulder), 208–14.

Vic-Ozouf Marignier, M. (1989) *La formation des départements et la représentation du territoire français à la fin du XVIIIe. siècle* (Paris).

Vicens Vives, J. (1940) *España: Geopolítica del Estado y del Imperio* (Madrid).

Vidal de la Blache, P. (1903) *Tableau de la Géographie de la France* (Paris).

Vidal de la Blache, P. (1910) 'Régions françaises', *Revue de Paris* 17, 6: 821–49.

Vidal de la Blache, P. (1917) *La France de l'Est* (Paris).

Vogt, E.A. (1996) '*Civilisation* and *Kultur*: Keywords in the History of French and German Citizenship', *Ecumene* 3, 2: 125–45.

von Bernhardi, F. (1914) *Germany and the Next War* (London).

von Raumer, K. (1953) *Ewiger Friede: Friedensrufe und Friedenspläne seit der Renaissance* (Munich).

von Treitschkte, H. (1916) *Politics* 2 vols (London).

Vovelle, M. (1993) *Le découverte de la politique: géopolitique de la Révolution française* (Paris).

Voyenne, B. (1964) *Histoire de l'idée européenne* (Paris).

Waever, O. (1990) 'Three Competing Europes: German, French, Russian', *International Affairs* 66, 3: 477–94.

Waever, O. (1993) 'Europe since 1945: Crisis to Renewal', in K. Wilson and J. van der Dussen (eds), *The History of the Idea of Europe* (London), 151–214.

Wallerstein, I. (1974) *The Modern World System: Capitalist Agriculture and the Origins of the European World-Economy in the Sixteenth Century* (London).

Wallerstein, I. (1980) *The Modern World System II: Mercantilism and the Consolidation of the European World-Economy 1600–1750* (London).

Wallerstein, I. (1991) *Geopolitics and Geoculture: Essays on the Changing World-System* (Cambridge).

Wandycz, P. (1992) *The Price of Freedom: A History of East Central Europe from the Middle Ages to the Present* (London).

Wanklyn, H.G. (1940) 'Geographical Aspects of Jewish Settlement East of Germany', *Geographical Journal* 95: 175–90.

Wanklyn, H.G. (1941) *The Eastern Marchlands of Europe* (London).

Wanklyn, H.G. (1961) *Friedrich Ratzel* (Cambridge).

Webb, W.P. (1952) *The Great Frontier* (Boston).

Weber, E. (1986a) 'L'Hexagon', in P. Nora (ed.), *Les lieux de mémoire*. Vol. 2, Part 2. *La nation* (Paris), 96–116.

Weber, E. (1986b) *France, Fin de Siècle* (Cambridge, MA).

Weigall, D. (1990) 'British Ideas of European Unity and Regional Confederation in the Context of Anglo–Soviet Relations, 1941–45', in M.L. Smith and P.M.R. Stirk (eds), *Making the New Europe: European Unity and the Second World War* (London), 156–68.

Weigert, H.W. (1941) *German Geopolitics* (Oxford).

Weigert, H.W. (1942) *Generals and Geographers: The Twilight of Geopolitics* (Oxford).

Weindling, P. (1989) *Health, Race and German Politics between National Unification and Nazism, 1870–1945* (Cambridge).

Wells, H.G. (1908) *New Worlds for Old* (London).

Whalen, R.W. (1984) *Bitter Wounds: German Victims of the Great War 1914–1939* (Ithaca).

White, H. (1973) *Metahistory: the Historical Imagination in Nineteenth-Century Europe* (Baltimore).

White, R. (1989) 'The Europeanism of Coudenhove-Kalergi', in P.M.R. Stirk (ed.), *European Unity in Context: The Interwar Period* (London), 23–40.

White, R. (1991) 'Cordial Caution: The British Response to the French Proposal for European Federal Union of 1930', in A. Bosco (ed.), *The Federal Idea* Vol. 1 (London), 236–62.

Whittlesey, D. (1940) 'A Utopia for Europe', *The New Republic* 12, February: 35–41

Wiedemer, P. (1993) 'The Idea Behind Coudenhove-Kalergi's Pan-European Union', *History of European Ideas* 16, 4: 827–33.

Wieviorka, M. (1994) 'Racism in Europe: Unity and Diversity', in A. Rattansi and S. Westwood (eds), *Racism, Modernity and Identity on the Western Front* (Cambridge), 173–88.

Wildes, A. (1977) *William Penn* (London).

Wilford, R.A. (1980) 'The Federal Union Campaign', *European Studies Quarterly* 10: 101–14.

Wilkinson, H.R. (1951) *Maps and Politics: A Review of the Ethnographic Cartography of Macedonia* (Liverpool).

Williams, A.M. (1991) *The European Community: The Contradictions of Integration* (Oxford).

Williams, C.H. (ed.) (1982) *National Separatism* (Cardiff).

Williams, C.H. (1989) 'The Question of National Congruence', in R.J. Johnston and P.J. Taylor (eds), *A World in Crisis? Geographical Perspectives* (Oxford), 229–65.

Williams, C.H. (ed.) (1993) *The Political Geography of the New World Order* (London).

Williams, R. (1976) *Keywords: A Vocabulary of Culture and Society* (London).

Wilson, K. and van der Dussen, J. (eds) (1993) *The History of the Idea of Europe* (London).

Wilson, W. (1919) *The State: Elements of Historical and Practical Politics* (London).

Winter, J.M. (1988) *The Experience of World War I* (London).

Winter, J.M. (1995) *Sites of Memory, Sites of Mourning: The Great War in European Cultural History* (Cambridge).

Wintle, M. (1996) 'Europe's Image: Visual Representations of Europe from the Earliest Times to the Twentieth Century', in M. Wintle (ed.), *Culture and Identity in Europe: Perceptions of Divergence and Unity in Past and Present* (Aldershot), 52–97.

Wintle, M. (1999) 'Renaissance Maps and the Construction of the Idea of Europe', *Journal of Historical Geography* 25, in press.

Wistrich, E. (1994) *The United States of Europe* (London).

Wittfogel, K. (1985) [1929] 'Geopolitics, Geographical Materialism and Marxism', *Antipode* 17: 21–72.

Wokler, R. (1995) *Rousseau* (London).

Wolf, E.R. (1982) *Europe and the People Without History* (Berkeley and Los Angeles).

Wolff, L. (1994) *Inventing Eastern Europe: The Map of Civilization on the Mind of the Enlightenment* (Stanford).

Woods, H.C. (1919) *The Cradle of the War: The Near East and Pan-Germanism* (London).

Woods, R. (1990) *A Changing of the Guard: Anglo–American Relations, 1941–1946* (Chapel Hill, NC).

Woods, R. and Jones, H. (1991) *Dawning of the Cold War: The United States' Quest for Order* (Athens, GA).

Woodward, D. (1985) 'Reality, Symbolism, Time and Space in Medieval World Maps', *Annals of the Association of American Geographers* 75: 510–21.

Woolf, S. (1989a) 'Statistics and the Modern State', *Comparative Studies in Society and History* 31, 3: 588–604.

Woolf, S. (1989b), 'French Civilization and Ethnicity in the Napoleonic Empire', *Past and Present* 124: 96–106.

Woolf, S. (1991) *Napoleon's Integration of Europe* (London).

Woolf, S. (1992) 'The Construction of a European World-View in the Revolutionary-Napoleonic Years', *Past and Present* 137: 72–101.

Wright, J. and Stafford, P. (1988) 'A Blueprint for World War? Hitler and the Hossbach Memorandum', *History Today* 38, 3: 11–17.

Wright, M. (1975) *The Theory and Practice of the Balance of Power, 1486–1914* (London).

Yergin, D. (1978) *Shattered Peace: The Origins of the Cold War and the National Security State* (London).

Young, J. (1993) *The Texture of Memory: Holocaust Memorials and Meaning* (New Haven).

Young, J.W. (1985a), 'Churchill's "No" to Europe: The "Rejection" of European Union by Churchill's Post-War Government, 1951-2', *Historical Journal* 28: 923–37.

Young, J.W. (1985b), *Britain, France and the Unity of Europe* (Leicester).

Young, J.W. (1991) *Cold War Europe, 1945–89: A Political History* (London).

Zeman, Z.A.B. (1991) *The Making and Breaking of Communist Europe* (Oxford).

Index

Note: Entries are derived from references in main text and titles of illustrations only